Communications and Control Engineering

Published titles include:

Randomized Algorithms for Analysis and Control of Uncertain Systems
Roberto Tempo, Giuseppe Calafiore and Fabrizio Dabbene

Stability and Stabilization of Infinite Dimensional Systems with Applications
Zheng-Hua Luo, Bao-Zhu Guo and Omer Morgul

Nonsmooth Mechanics (Second edition)
Bernard Brogliato

Nonlinear Control Systems II
Alberto Isidori

L_2-Gain and Passivity Techniques in nonlinear Control
Arjan van der Schaft

Control of Linear Systems with Regulation and Input Constraints
Ali Saberi, Anton A. Stoorvogel and Peddapullaiah Sannuti

Robust and H∞ Control
Ben M. Chen

Computer Controlled Systems
Efim N. Rosenwasser and Bernhard P. Lampe

Dissipative Systems Analysis and Control
Rogelio Lozano, Bernard Brogliato, Olav Egeland and Bernhard Maschke

Control of Complex and Uncertain Systems
Stanislav V. Emelyanov and Sergey K. Korovin

Robust Control Design Using H∞ Methods
Ian R. Petersen, Valery A. Ugrinovski and Andrey V. Savkin

Model Reduction for Control System Design
Goro Obinata and Brian D.O. Anderson

Control Theory for Linear Systems
Harry L. Trentelman, Anton Stoorvogel and Malo Hautus

Functional Adaptive Control
Simon G. Fabri and Visakan Kadirkamanathan

Positive 1D and 2D Systems
Tadeusz Kaczorek

Identification and Control Using Volterra Models
F.J. Doyle III, R.K. Pearson and B.A. Ogunnaike

Non-linear Control for Underactuated Mechanical Systems
Isabelle Fantoni and Rogelio Lozano

Robust Control (Second edition)
Jürgen Ackermann

Flow Control by Feedback
Ole Morten Aamo and Miroslav Krstiæ

Learning and Generalization (Second edition)
Mathukumalli Vidyasagar

Constrained Control and Estimation
Graham C. Goodwin, María M. Seron and José A. De Doná

Randomized Algorithms for Analysis and Control of Uncertain Systems
Roberto Tempo, Giuseppe Calafiore and Fabrizio Dabbene

Switched Linear Systems
Zhendong Sun and Shuzhi S. Ge

Andrea Bacciotti and Lionel Rosier

Liapunov Functions and Stability in Control Theory

2nd Edition

With 20 Figures

Andrea Bacciotti, Prof. Dr.
Politecnico Torino, Dipto. Matematica, Corso Duca Degli Abruzzi 24,
10129 Torino, Italy

Lionel Rosier, Prof.
Institut Elie Cartan
Université Nancy 1
B.P. 239
54506 Vandœuvre-lès-Nancy Cedex, France

Series Editors
E.D. Sontag · M. Thoma · A. Isidori · J.H. van Schuppen

Originally published as volume 267 in the series "Lecture Notes in Control and Information Sciences,"
Springer London (2001)

ISSN 0178-5354

ISBN 10 3-540-21332-5 Springer Berlin Heidelberg New York
ISBN 13 978-3-540-21332-1 Springer Berlin Heidelberg New York

Library of Congress Control Number: 2005921904

This work is subject to copyright. All rights are reserved, whether the whole or part of the material is concerned, specifically the rights of translation, reprinting, reuse of illustrations, recitation, broadcasting, reproduction on microfilm or in other ways, and storage in data banks. Duplication of this publication or parts thereof is permitted only under the provisions of the German Copyright Law of September 9, 1965, in its current version, and permission for use must always be obtained from Springer-Verlag. Violations are liable to prosecution under German Copyright Law.

Springer is a part of Springer Science+Business Media
springeronline.com
© Springer-Verlag Berlin Heidelberg 2005
Printed in The Netherlands

The use of general descriptive names, registered names, trademarks, etc. in this publication does not imply, even in the absence of a specific statement, that such names are exempt from the relevant protective laws and regulations and therefore free for general use.

Typesetting: Data conversion by author.
Final processing: PTP-Berlin Protago-T$_E$X-Production GmbH, Germany
Cover-Design: PTP-Berlin Protago-T$_E$X-Production GmbH, Germany
Printed on acid-free paper   89/3141/Yu - 5 4 3 2 1 0

Preface

We are interested in mathematical models of input systems, described by continuous-time, finite dimensional ordinary differential equations

$$\dot{x} = f(t, x, u) \tag{1}$$

where $t \geq 0$, $x = (x_1, \ldots, x_n) \in \mathbb{R}^n$ represents the state variables, $u = (u_1, \ldots, u_m) \in \mathbb{R}^m$ represents the input variables and $f = (f_1, \ldots, f_n) : [0, +\infty) \times \mathbb{R}^n \times \mathbb{R}^m \to \mathbb{R}^n$. Together with (1), we will often consider the *unforced associated system*

$$\dot{x} = f(t, x, 0) \ . \tag{2}$$

Basically, (2) accounts for the "internal" behavior of the system. More precisely, (2) describes the natural dynamics of (1) when no energy is supplied through the input channels. The analysis of the "external" behavior is rather concerned with the effect of the inputs (disturbances or exogenous signals) on the evolution of the state response of (1).

Physical systems are usually expected to exhibit a "stable" behavior. A primary aim of this book is to survey some possible mathematical definitions of internal and external stability in a nonlinear context and to discuss their characterizations in the framework of the Liapunov functions method.

We will also consider the problem of achieving a more desirable stability behavior (both from the internal and the external point of view) by means of properly designed feedback laws. To this end, it is convenient to think of the input as a sum $u = u_e + u_c$. The term u_e represents external forces, while u_c is actually available for control action. Roughly speaking, (1) is said to be "stabilizable" if there exists a map $u_c = k(t, x)$ such that the closed loop system

$$\dot{x} = f(t, x, k(t, x) + u_e) \tag{3}$$

exhibits improved (internal and/or external) stability performances.

Intimate relationships among all these aspects of systems analysis emerge with some evidences from classical linear systems theory. In particular, as we shall see at the beginning of Chapter 2, the external behavior of a linear system is strongly related to its internal structure. On the contrary, dealing with nonlinear systems these connections become weaker and need a more delicate treatment.

We shall see in particular that the approach to stability and stabilizability of nonlinear systems rests much more heavily on the method of Liapunov functions. Thus, we are led to emphasize the interest in a variety of theorems which state, under minimal assumptions, the existence of Liapunov functions with suitable properties. These theorems are usually called "converse Liapunov theorems". A secondary aim of this book is to illustrate the state of the art on this subject, and to present some recent developments.

We have not yet specified what kind of assumptions should be made about the map f which appears at the right hand side of (1) and about the admissible inputs.

The class of admissible inputs should be so large to include representations of all signals commonly used in engineering applications. To this purpose, it is well known that in certain circumstances, a discontinuous function often is more suited than a continuous one. Thus, throughout these notes, we shall adopt the following agreement:

(**I**) the class of *admissible inputs* is constituted by all measurable, essentially bounded functions $u : [0, +\infty) \to \mathbb{R}^m$.

To establish the assumptions about f is a more delicate task. In a classical "smooth" setting, it seems natural to ask that f is time invariant, namely $f(t, x, u) = f(x, u)$, and at least continuous as a function of x, u, though additional regularity could be required for certain purposes[1]. This is actually the point of view we intend to adopt at the beginning but, as long as we proceed in our exposition, it will become clear that the smooth setting is too conservative for certain developments. This occurs in particular when we seek Liapunov functions of (Liapunov or Lagrange) stable systems or when we aim to design internally asymptotically stabilizing feedback laws. Indeed, the solution of

---

[1] Recent results of the so-called geometric control theory apply to systems whose right hand side can be represented as a family of C^∞ or real analytic vector fields (see [79], [80], [155]).

these problems cannot be found in general within a pre-assigned class of time invariant smooth functions, unless severe restrictions are made on the system under consideration. We will be so led to introduce in our treatment nondifferentiable functions and differential equations with discontinuous right hand side.

We remark that differential equations with discontinuous right hand side arise in many engineering and physical applications. Historically, one of the main motivation was the study of the motion of a body with one degree of freedom subject to an elastic force, in presence of both viscous and dry friction ([58], [54]). This is modelled by the second order equation

$$\ddot{x} + kx + b\dot{x} + a\,\mathrm{sgn}\,\dot{x} = 0$$

or, equivalently, by the two dimensional system

$$\begin{cases} \dot{x} = y \\ \dot{y} = -kx - by - a\,\mathrm{sgn}\,y \end{cases} \quad (4)$$

(here, k, b and a are positive constants). For $y \neq 0$, the motion is correctly represented by the solutions of the system. But if the body reaches a position $(x, 0)$ with $-\frac{a}{k} < x < \frac{a}{k}$, our intuition suggests that the elastic force is too weak. It cannot overcome the dry friction, and the body remains at rest. This intuition is easily confirmed by physical observation, but it is not reflected by system (4), at least as far as the solutions are intended in the usual sense.

Differential equations with discontinuous right hand side play an important role also in *variable structure control* methodologies. Consider, for simplicity, a time-invariant system

$$\dot{x} = f(x, u) \;.$$

In variable structure control theory, the goal is to track a path lying on a hypersurface Σ defined by an equation $s(x) = 0$, where $s(x)$ is a smooth function. To this purpose, it is often convenient to use discontinuous feedback, say for instance

$$u = k(x) = \begin{cases} 1 & \text{if } s(x) > 0 \\ -1 & \text{if } s(x) < 0 \end{cases}.$$

Clearly, the closed loop system

$$\dot{x} = f(x, k(x))$$

turns out to be discontinuous even if $f(x,u)$ is a smooth function. The desired motion is given by a trajectory sliding on Σ; in general, it is not a solution in the usual sense of the closed loop system.

We finally remark that discontinuities of the velocity and sometimes also of the state evolution are a typical feature of the so-called *hybrid dynamical systems*: the book [152] provides a nice introduction on this subject, with many practical examples (manual transmission, temperature control, electric circuits with diodes, and many others).

These remarks point out that the treatment of differential equations with discontinuous right hand side requires a generalization of the classical notion of solution.

To be prepared for this extension, in Chapter 1 we recall some preliminary material about existence of solutions for ordinary differential equations and differential inclusions.

The main subject will be addressed starting from Chapter 2. As already mentioned, in Chapter 2 we focus more precisely on the case where the right hand side of (1) is time invariant and continuous with respect to both x, u. The reason why we prefer to begin with such a restricted class of systems is twofold. First, the more general approach could be felt at that point unmotivated and too abstract. Second, the main notions, methods and achievements available in the literature about stability and stabilizability theory of control systems have been mostly obtained, in the last few years, just for this class of systems. Of course, the choice of proceeding from the simplest situation to the more general one, implies also a few of complications (for instance, the need of a progressive updating of definitions and results when we shall undertake certain extensions) but gives a clearer perspective of problems and theoretical difficulties.

A first attempt to re-interpret our problems in a more general context is made in Chapter 3, where we consider time varying systems. We focus in particular on possible notions of internal stability and on their relationships. Although we are able to give some more precise results about existence of Liapunov functions and of stabilizing feedback, we shall see that the picture of the situation is not yet completely satisfactory.

The goal of replacing the classical smooth setting by a more general time dependent and "nonsmooth" one, will be fully pursued in Chapter 4, where we finally consider systems of the general form (1), and f is allowed to be discontinuous with respect to x. More precisely, in Chapter 4 we discuss direct and converse theorems about stability and asymptotic stability, together with their applications to external stabilization. We present also a new approach which allows us to prove in a unified manner several recent results. The proof given here is considerably shorter and easier than other proofs available in the original papers.

Certain additional properties of Liapunov functions will be discussed in Chapter 5. Here, we consider again the case of systems of ordinary differential equations, with time invariant and smooth right hand side. The topics include existence of analytic or homogeneous Liapunov functions and their symmetries, and relationship between Liapunov functions and decay of trajectories.

Finally, in Chapter 6 we review some tools from nonsmooth analysis which can be useful in the investigation of nondifferentiable systems with discontinuous Liapunov functions.

Contents

Preface V

1 **Differential equations** 1
   1.1 Recall about existence results 2
   1.2 Differential inclusions 4
       1.2.1 The upper semi-continuous case 4
       1.2.2 The Lipschitz continuous case 8
   1.3 Appendix 13

2 **Time invariant systems** 19
   2.1 The linear case 19
       2.1.1 Stability 19
       2.1.2 Internal stabilization 22
       2.1.3 External stabilization 23
       2.1.4 Quadratic forms 24
   2.2 Nonlinear systems: stability 27
       2.2.1 Internal notions 27
       2.2.2 Converse theorems 30
       2.2.3 Generalized Liapunov functions 37
       2.2.4 Absolute stability 39
       2.2.5 Stability and robustness 44
       2.2.6 The Invariance Principle 47
       2.2.7 The domain of attraction 49
       2.2.8 Comparison functions 55
       2.2.9 External notions 58
   2.3 Nonlinear systems: stabilization 62
       2.3.1 Necessary condition for internal stabilization 63
       2.3.2 Asymptotic controllability and local controllability . . . 68

		2.3.3	Affine systems: internal stabilization	70
		2.3.4	Affine systems: external stabilization	77
	2.4	Output systems		79
	2.5	Cascade systems		80
	2.6	Appendix		83

3 Time varying systems — 89
    3.1 Two examples 89
    3.2 Reformulation of the basic definitions 94
          3.2.1 Stability and attraction 94
          3.2.2 Time dependent Liapunov functions 101
    3.3 Sufficient conditions 104
    3.4 Converse theorems 105
          3.4.1 Asymptotic stability 105
          3.4.2 Uniform stability 105
    3.5 Robust stability 108
    3.6 Lagrange stability 112
    3.7 Discontinuous right hand side 113
    3.8 Time varying feedback 113

4 Differential inclusions — 119
    4.1 Global asymptotic stability 119
          4.1.1 Sufficient conditions 119
          4.1.2 Time invariant systems 121
          4.1.3 Time varying systems 122
          4.1.4 Proof of the converse of second Liapunov theorem . . . 125
    4.2 Robust stability 148
          4.2.1 Sufficient conditions 148
          4.2.2 Converse of first Liapunov theorem 152
          4.2.3 Proof of the converse of first Liapunov theorem 153
          4.2.4 Application to external stabilization 160
    4.3 Nonsmooth Liapunov functions 161

5 Additional properties of strict Liapunov functions — 167
    5.1 Estimates for the convergence of trajectories 170
          5.1.1 Exponential stability 170
          5.1.2 Rational stability 171
          5.1.3 Finite-time stability 175
    5.2 Analyticity 176

	5.2.1	Analytic unsolvability of the stability problem	176
	5.2.2	Analytic Liapunov functions	177
	5.2.3	Holomorphic systems	180
5.3	Weighted homogeneity		180
	5.3.1	A few definitions	181
	5.3.2	Homogeneous Liapunov functions	183
	5.3.3	Application to the stabilization problem	186
5.4	Symmetries and Liapunov functions		192
	5.4.1	Discrete symmetry	193
	5.4.2	Infinitesimal symmetry	193
	5.4.3	Symmetric Liapunov functions	196

6 Monotonicity and generalized derivatives   203

6.1	Tools from nonsmooth analysis	203
6.2	Functions of one variable	207
6.3	Ordinary differential equations	209
6.4	Differential inclusions	212
6.5	Monotonicity and the proximal gradient	215

Bibliography   219

Index   233

List of Abbreviations   237

Chapter 1

Differential equations

In what follows, $\mathbb{N}, \mathbb{Z}, \mathbb{Q}, \mathbb{R}$ represent respectively the sets of natural, integer, rational and real numbers. Sometimes, we may use also the notation $\mathbb{R}^+ = [0, +\infty)$ and $\mathbb{N}^* = \mathbb{N} \setminus \{0\}$.

Let $N \in \mathbb{N}^*$. The norm of a vector $v = (v_1, \ldots, v_N) \in \mathbb{R}^N$ is denoted by $||v||$. As is well known, for finite dimensional vector spaces all the norms are equivalent. Actually, the choice of the norm does not matter in the first three chapters. However, in view of the developments of Chapter 4, it is convenient to take the sup-norm

$$||v|| = \max\{|v_i| : 1 \leq i \leq N\}.$$

The Hausdorff distance between nonempty, compact subsets of $\mathbb{R}^N$ will be denoted by h. We recall that

$$h(A, B) = \max\{\sup_{a \in A} \operatorname{dist}(a, B), \sup_{b \in B} \operatorname{dist}(b, A)\}$$

where $\operatorname{dist}(a, B) = \inf_{b \in B} ||a - b||$.

For $x \in \mathbb{R}^N$ and $r > 0$, the open ball of center x and radius r is denoted by

$$B_r(x) = \{y \in \mathbb{R}^N : ||y - x|| < r\}.$$

Of course, $\overline{B_r(x)}$ denotes the closed ball. When $x = 0$, we shall write simply B_r instead of $B_r(0)$. We shall also use the symbol B^r for the complement of a closed ball, namely

$$B^r = \{y \in \mathbb{R}^N : ||y|| > r\} = \mathbb{R}^N \setminus \overline{B_r}.$$

Finally, let $g : \Omega \to \mathbb{R}^M$, where $\Omega \subseteq \mathbb{R}^N$. The function g is said to be *locally Lipschitz continuous* on Ω if for each $\bar{x} \in \Omega$ there exist positive real numbers L, δ such that

$$x', x'' \in B_\delta(\bar{x}) \cap \Omega \implies ||g(x') - g(x'')|| \leq L||x' - x''|| \ .$$

1.1 Recall about existence results

The first natural question about a system of the form (1) concerns of course the existence of (local) solutions corresponding to any admissible input. Throughout this chapter we assume that $u(t)$ is fixed, so that we can adopt the simplified notation $f(t, x) = f(t, x, u(t))$. We are therefore led to consider a system of ordinary differential equations of the form

$$\dot{x} = f(t, x) \tag{1.1}$$

where $f(t, x)$ is defined for all $x \in \mathbb{R}^n$ and $t \geq 0$. As is well known, Peano's Theorem states that if $f(t, x)$ is continuous on $[0, +\infty) \times \mathbb{R}^n$, then for each initial pair $(t_0, x_0) \in [0, +\infty) \times \mathbb{R}^n$ there exists at least one local *classical solution* $x(t) : I \to \mathbb{R}^n$ such that $x(t_0) = x_0$. Here, I is an interval of real numbers such that $t_0 \in I \subseteq [0, +\infty)$. The qualifier "classical" emphasizes that $x(t)$ is of class C^1 and

$$\dot{x}(t) = f(t, x(t)) \qquad \forall t \in I \ .$$

The continuity assumption required by Peano's Theorem is too restrictive for applications to control theory. Indeed, in general admissible inputs are assumed to be only measurable and essentially bounded. Therefore, even if the right hand side of (1) is continuous, we cannot hope that the resulting map $f(t, x) = f(t, x, u(t))$ is continuous.

The following set of assumptions for (1.1) seems to be more appropriate:

(**A$_1$**) the function $f(t, x)$ is locally essentially bounded on $[0, +\infty) \times \mathbb{R}^n$

(**A$_2$**) for each $x \in \mathbb{R}^n$, the function $t \mapsto f(t, x)$ is measurable

(**A$_3$**) for a.e. $t \geq 0$, the function $x \mapsto f(t, x)$ is continuous.

A function $x(t)$ is called a local *Carathéodory solution* of (1.1) on the interval I if it is absolutely continuous on every compact subinterval of I and satisfies

$$\dot{x}(t) = f(t, x(t)) \quad \text{a.e. } t \in I.$$

Carathéodory's Theorem states that if assumptions ($\mathbf{A}_1$), ($\mathbf{A}_2$), ($\mathbf{A}_3$) are fulfilled, then for each initial pair $(t_0, x_0) \in [0, +\infty) \times \mathbb{R}^n$ there exists an interval I with $t_0 \in I$ and a Carathéodory solution $x(t)$ defined on I.

For a system of the form (1.1), the set of all local Carathéodory solutions corresponding to a given initial pair (t_0, x_0) will be denoted by $\mathcal{S}_{t_0, x_0}$. When we need to emphasize the dependence of a particular solution $x(t) \in \mathcal{S}_{t_0, x_0}$ on the initial time and state, we shall use the notation $x(t) = x(t; t_0, x_0)$.

Moreover, when (1.1) results from an input system like (0.1) and we want to emphasize the dependence of solutions on the input $u(t)$, we shall write respectively $\mathcal{S}_{t_0, x_0, u(\cdot)}$ and $x(t) = x(t; t_0, x_0, u(\cdot))$.

Remark 1.1 Of course, any classical solution is also a Carathéodory solution. To show that the converse is false, consider the following simple one-dimensional equation

$$\dot{x} = f(t, x) = a(t)x$$

where

$$a(t) = \begin{cases} 0 & \text{if } t \in \mathbb{Q} \\ 1 & \text{if } t \in \mathbb{R} \setminus \mathbb{Q} . \end{cases}$$

For each initial pair $(0, x_0)$ with $x_0 \neq 0$, the set of classical solutions is empty, but there is a Carathéodory solution of the form $x = e^t x_0$. ∎

Peano's and Carathéodory's Theorems only guarantee in general the existence of local solutions. A typical additional assumption is *local Lipschitz continuity* with respect to x:

($\mathbf{A}_4$) for each point $(\bar{t}, \bar{x}) \in [0, +\infty) \times \mathbb{R}^n$ there exist $\delta > 0$ and a positive function $l(t) : [0, +\infty) \to \mathbb{R}$ such that $l(t)$ is locally integrable and

$$\|f(t, x') - f(t, x'')\| \leq l(t) \|x' - x''\|$$

for each t, x' and x'' such that $|t - \bar{t}| \leq \delta$, $\|x' - \bar{x}\| \leq \delta$ and $\|x'' - \bar{x}\| \leq \delta$.

Under the assumptions ($\mathbf{A}_1$), ($\mathbf{A}_2$), ($\mathbf{A}_3$) and ($\mathbf{A}_4$), it is possible to prove local uniqueness and continuity of solutions with respect to the initial data. In particular, the following holds.

(C) let $(\bar{t}, \bar{x}) \in [0, +\infty) \times \mathbb{R}^n$ be fixed, and assume that $x(t; \bar{t}, \bar{x})$ is defined on some closed interval $[\alpha, \beta]$ (with $\alpha \leq \bar{t} \leq \beta$). Then for each $\varepsilon > 0$ there exists $\delta > 0$ such that for each pair (τ, ξ) with

$$|\tau - \bar{t}| < \delta , \qquad \|\xi - \bar{x}\| < \delta$$

the solution $x(t; \tau, \xi)$ is defined for $\alpha \leq t \leq \beta$ and

$$\|x(t; \tau, \xi) - x(t; \bar{t}, \bar{x})\| < \varepsilon$$

for each $t \in [\alpha, \beta]$.

These and other results about ordinary differential equations can be found in many usual textbooks (see for instance [125], [70], [60]).

1.2 Differential inclusions

In this section we illustrate how differential inclusions arise in the mathematical theory of control systems. Moreover, we recall the main existence results needed in the following chapters. In particular, we show that the existence of Filippov solutions for discontinuous differential equations can be actually deduced from an existence theorem for differential inclusions.

1.2.1 The upper semi-continuous case

As already mentioned in the Introduction, for certain applications of control theory we need to resort to differential equations whose right hand side is discontinuous not only with respect to t, but also with respect to the state variable x. Indeed, even if the system is modeled by smooth vector fields, discontinuities may be inevitably introduced when closed loop solutions of certain problems are required.

Note that if the right hand side of (1.1) is not continuous with respect to x, then the usual notions of solution (classical or Carathéodory) do not apply. The more common way to overcome the difficulty is to replace (1.1) by a differential inclusion of the form

$$\dot{x} \in F(t, x) . \tag{1.2}$$

A *solution* of (1.2) is any function $x(t)$ defined on some interval $I \subseteq [0, +\infty)$ which is absolutely continuous on each compact subinterval of I and such that

$$\dot{x}(t) \in F(t, x(t)) \qquad \text{a.e. on } I.$$

Then, by definition, a function $x(t)$ is a *generalized solution* of (1.1) if and only if it is a solution of the associated differential inclusion (1.2).

Of course, to give a precise meaning to the notion of generalized solution we need to assign a rule which enables us to associate a set valued map $F(t,x)$ to the discontinuous map $f(t,x)$. This can be done in many way. In this book, we adopt the approach due to A.F. Filippov ([60], [58]), which is based on the following idea. Set

$$F(t,x) = \mathbf{K}_x f(t,x) \underset{\text{def}}{=} \bigcap_{\delta>0} \bigcap_{\mu(N)=0} \overline{\text{co}} \left\{ f(t, B_\delta(x) \backslash N) \right\} \qquad (1.3)$$

where $\overline{\text{co}}$ denotes the convex closure of a set and μ is the usual Lebesgue measure of $\mathbb{R}^n$. The generalized solutions of (1.1) defined according to (1.2), (1.3) will be called *Filippov solutions*. When $f(t,x)$ is (Lebesgue) measurable and locally bounded, there is also an equivalent (perhaps more intuitive) definition (see [104]). Indeed, it is possible to prove that for each $t \geq 0$, there exists a set $N_0^t \subset \mathbb{R}^n$ (depending on f and t) with $\mu(N_0^t) = 0$ such that, for each $N \subset \mathbb{R}^n$ with $\mu(N) = 0$, and for each $x \in \mathbb{R}^n$,

$$\mathbf{K}_x f(t,x) = \text{co}\{v : \exists \{x_i\} \text{ with } x_i \to x$$
$$\text{such that } x_i \notin N_0^t \cup N \text{ and } v = \lim f(t, x_i)\}. \qquad (1.4)$$

In [104], the reader will find also some useful rules of calculus for the "operator" $\mathbf{K}_x$.

Remark 1.2 Strictly speaking, Filippov solutions are not a generalization of classical solutions. Consider for instance the scalar equation

$$\dot{x} = f(x)$$

defined by

$$f(x) = \begin{cases} 1 & \text{if } x \neq 0 \\ 0 & \text{if } x = 0. \end{cases}$$

In the classical sense, the constant function $x(t) \equiv 0$ is a solution. However, it is not a solution in Filippov's sense. On the other hand, if

$$f(x) = \begin{cases} 1 & \text{if } x \geq 0 \\ -1 & \text{if } x < 0 \end{cases}$$

then $x(t) \equiv 0$ is a Filippov solution but not a classical one. ∎

The set valued map $F(t,x)$ associated to a discontinuous equation by virtue of construction (1.3) has a number of nice properties. The first result is more or less evident.

Proposition 1.1 *Assume that $f(t,x)$ is locally bounded on $[0,+\infty) \times \mathbb{R}^n$. Then, the set valued map $F(t,x) = \mathbf{K}_x f(t,x)$ is defined for each pair $(t,x) \in [0,+\infty) \times \mathbb{R}^n$ and its values are nonempty, compact and convex. Moreover, it is locally bounded on $[0,+\infty) \times \mathbb{R}^n$ that is, for each compact set $K \subset [0,+\infty) \times \mathbb{R}^n$ there exists $M > 0$ such that*

$$F(t,x) \subset B_M$$

for each $(t,x) \in K$.

From now on, we limit ourselves to compact valued maps.

Definition 1.1 *Let $\mathcal{A}$ be a σ-algebra of subsets of $\mathbb{R}^{n_1}$. A set valued map G from $\mathbb{R}^{n_1}$ to $\mathbb{R}^{n_2}$ is said to be $\mathcal{A}$-measurable if for each open set $\Omega \subseteq \mathbb{R}^{n_2}$, the set*

$$\{x \in \mathbb{R}^{n_1} : G(x) \cap \Omega \neq \emptyset\}$$

belongs to $\mathcal{A}$. A set valued map G from $\mathbb{R}^{n_1}$ to $\mathbb{R}^{n_2}$ with compact values is upper semi-continuous if for each x_0 and for each $\varepsilon > 0$ there exists $\delta > 0$ such that

$$G(x) \subseteq G(x_0) + B_\varepsilon$$

provided that $x \in B_\delta(x_0)$.

Note that if G is single valued, these definitions agree with the usual definitions of $\mathcal{A}$-measurable and, respectively, continuous function.

Theorem 1.1 *Let $f(t,x)$ be locally bounded and let $F(t,x) = \mathbf{K}_x f(t,x)$. Then, $F(t,x)$, as a set valued map of x, is upper semi-continuous for each t.*

Theorem 1.2 *Let $f(t,x)$ be locally bounded and let $F(t,x) = \mathbf{K}_x f(t,x)$. If $f(t,x)$ is Lebesgue measurable (respectively, Borel measurable), then $F(t,x)$ is Lebesgue measurable (respectively, Borel measurable).*

It follows from Theorem 1.2 that if $f(t,x)$ is a locally bounded and Borel measurable map, then $F(t,x)$, as a function of t, is a Borel measurable map for every x, as well. Indeed for fixed $x = \bar{x}$, $t \mapsto F(t,\bar{x})$ can be seen as the composite map of $t \mapsto (t,\bar{x})$ and $(t,x) \mapsto F(t,x)$. Hence, we conclude in particular that:

if $f(t,x)$ is a locally bounded and Borel measurable map,
then $t \mapsto F(t,x)$ is Lebesgue measurable for each x.

Clearly, the same reasoning does not work if $f(t,x)$ is assumed to be only Lebesgue measurable. Nevertheless, the conclusion remains true.

Theorem 1.3 *Let $f(t,x)$ be locally bounded and Lebesgue measurable with respect to (t,x). Then, $F(t,x) = \mathbf{K}_x f(t,x)$ is Lebesgue measurable as a function of t, for each fixed x.*

Proposition 1.1 and Theorem 1.1 can be proven as in [7] p. 102; the proofs of Theorems 1.2 and 1.3 are given in the Appendix of this chapter.

Summing up, Proposition 1.1, Theorem 1.1 and Theorem 1.3 together show that if $f(t,x)$ is locally bounded and Lebesgue measurable with respect to the pair (t,x) on $[0,+\infty) \times \mathbb{R}^n$, then $F(t,x) = \mathbf{K}_x f(t,x)$ satisfies the following set of assumptions:

(H$_1$) $F(t,x)$ is a nonempty, compact, convex subset of $\mathbb{R}^n$, for each $t \geq 0$ and each $x \in \mathbb{R}^n$

(H$_2$) $F(t,x)$, as a set valued map of x, is upper semi-continuous for each $t \geq 0$

(H$_3$) $F(t,x)$, as a set valued map of t, is Lebesgue measurable for each $x \in \mathbb{R}^n$

(H$_4$) $F(t,x)$ is locally bounded.

The following theorem can be found for instance in [54], [60].

Theorem 1.4 *Let $F(t,x)$ be a set valued map which fulfills all the assumptions $(\mathbf{H}_1), \ldots, (\mathbf{H}_4)$. Then, for each pair $(t_0, x_0) \in [0, +\infty) \times \mathbb{R}^n$ there exist an interval I and at least a solution $x(t): I \to \mathbb{R}^n$ of (1.2) such that $t_0 \in I$ and*

$$x(t_0) = x_0 . \tag{1.5}$$

Remark 1.3 By virtue of Theorems 1.1 and 1.3, the previous statement provides also an existence result for Filippov solutions of the initial value problem

(1.1), (1.5) under the assumption that f is Lebesgue measurable (with respect to both variables) and locally bounded. Under the same assumptions, an existence theorem for solutions of discontinuous differential equations is obtained in [60], p. 85. However, in [60] the theorem is proved in a different way, without using Theorem 1.4.

1.2.2 The Lipschitz continuous case

In the previous subsection we discovered an important link between control theory and the theory of differential inclusions. A second, important link is given by the fact that a system with free inputs can be actually reviewed as a differential inclusion of a particular type. We need to recall some definitions.

Definition 1.2 *A set valued map G from $\mathbb{R}^{n_1}$ to $\mathbb{R}^{n_2}$ is Hausdorff continuous at a point $\bar{x} \in \mathbb{R}^{n_1}$ if for each $\varepsilon > 0$ there exists $\delta > 0$ such that $x \in B_\delta(\bar{x})$ implies $h(G(x), G(\bar{x})) < \varepsilon$.*

Definition 1.3 *A set valued map F from $[0, +\infty) \times \mathbb{R}^n$ to $\mathbb{R}^n$ is said to be locally Lipschitz continuous (in Hausdorff sense) with respect to x if for each point $(\bar{t}, \bar{x}) \in [0, +\infty) \times \mathbb{R}^n$ there exist $\delta > 0$ and a positive function $l(t) : [0, +\infty) \to \mathbb{R}$ such that $l(t)$ is locally integrable and*

$$h(F(t, x_1), F(t, x_2)) \leq l(t)\|x_1 - x_2\| \tag{1.6}$$

for each t, x_1 and x_2 such that $|t - \bar{t}| \leq \delta$, $\|x_1 - \bar{x}\| \leq \delta$ and $\|x_2 - \bar{x}\| \leq \delta$.

Consider now a system of the form (1) and assume that $f(t, x, u)$ is locally bounded, Lebesgue measurable with respect to t for each pair (x, u) and continuous with respect to x and u for each t[1]. Let U be a given subset of $\mathbb{R}^m$, and assume now that an input function $u(\cdot)$ is admissible only if it (is measurable, essentially bounded and, in addition) fulfills the constraint $u(t) \in U$, a.e. $t \in [0, +\infty)$. Then, it is evident that every solution of (1) corresponding to an admissible input is a solution of a differential inclusion (1.2) where the right hand side is now defined by

$$F(t, x) = f(t, x, U) \ .$$

A celebrated theorem by Filippov ([57]) states that the converse is also true, provided that
(∗) $f(t, x, u)$ is continuous and U is a compact set.

---
[1] This implies that for each admissible input $f(t, x, u(t))$ satisfies Carathéodory conditions ($\mathbf{A}_1$), ($\mathbf{A}_2$), ($\mathbf{A}_3$) ([60], Lemma 1 p. 3).

On the other hand, it is not difficult to show that if (∗) holds, then $F(t,x) = f(t,x,U)$ is Hausdorff continuous. If in addition $f(t,x,u)$ is locally Lipschitz continuous with respect to x (uniformly with respect to u) then $F(t,x)$ is also locally Lipschitz Hausdorff continuous with respect to x.

We are now ready to address the existence issue for the initial value problem (1.2), (1.5) under the alternative assumption that $F(t,x)$ is locally Lipschitz continuous with respect to x. The following theorem was proved in [59]: beside local Lipschitz continuity, the author requires that $F(t,x)$ is Hausdorff continuous, with nonempty compact values: in fact, the Hausdorff continuity assumption can be relaxed, by simply requiring that $F(t,x)$ is locally bounded and measurable with respect to t for each x ([21]).

Theorem 1.5 *Assume that $F(t,x)$ is locally Lipschitz continuous with respect to x on $[0, +\infty) \times \mathbb{R}^n$. Assume further that it has nonempty compact values and that (**H$_3$**), (**H$_4$**) hold. Let I be a bounded closed interval, $I \subset [0, +\infty)$ and let $\gamma(t) : I \to \mathbb{R}^n$ be an absolutely continuous function. Let $t_0 \in I$ and assume that*

$$\|\gamma(t_0) - x_0\| \leq r \tag{1.7}$$

and

$$\mathrm{dist}\,(\dot\gamma(t), F(t, \gamma(t))) \leq \rho(t) \tag{1.8}$$

for a.e. $t \in I$, where $r \geq 0$ and the function ρ is integrable on I. Then, there exist an integrable function $l(t) : I \to \mathbb{R}^+$, an interval $J \subseteq I$ ($t_0 \in J$) and a solution $x(t) : J \to \mathbb{R}^n$ of the initial value problem (1.2), (1.5) such that

$$\|x(t) - \gamma(t)\| \leq L(t) \tag{1.9}$$

for each $t \in J$, and

$$\|\dot x(t) - \dot\gamma(t)\| \leq l(t) L(t) + \rho(t) \tag{1.10}$$

for a.e. $t \in J$, where

$$L(t) = r e^{m(t)} + \left| \int_{t_0}^t e^{m(t) - m(s)} \rho(s)\, ds \right|, \qquad m(t) = \left| \int_{t_0}^t l(s)\, ds \right|.$$

Note that Theorem 1.5 provides an existence result (when applied, for instance, with $\gamma(t) = x_0$) and, at the same time, a result on continuous dependence of solutions with respect to initial data (when applied with $\rho \equiv 0$, in which case γ is itself a solution). Note also that no convexity assumption is needed.

For reader's convenience, we report a proof of Theorem 1.5.

Proof of Theorem 1.5. Let $b > r$, and let $l(t) : \bar{I} \to \mathbb{R}^+$ be an integrable function such that (1.6) holds for $t \in \bar{I}$ and $x_1, x_2 \in \overline{B_b(\gamma(t))}$. Such a function exists by virtue of the compactness argument. Moreover, let $J \subseteq I$ be such that $L(t) \leq b$ for each $t \in J$. Such an interval exists, since $L(t_0) = r$.

Let us construct a sequence of absolutely continuous functions $\{x_i(t)\}$ in the following way. First, we set $x_0(t) = \gamma(t)$, and recall that since $\gamma(t)$ is absolutely continuous, then $\dot{x}_0(t)$ is integrable on I. According to Lemma 1.2 of [72], there exists a measurable function $v_0(t)$ such that

$$v_0(t) \in F(t, x_0(t)) \quad \text{and} \quad ||v_0(t) - \dot{x}_0(t)|| = \text{dist}\,(\dot{x}_0(t), F(t, x_0(t))) \quad (1.11)$$

a.e. $t \in I$. Since $\rho(t)$ is integrable on I, it follows from (1.8) that $v_0(t)$ is integrable on I, as well. So, we can define

$$x_1(t) = x_0 + \int_{t_0}^{t} v_0(s)\,ds . \quad (1.12)$$

Using (1.11) and (1.7), we have

$$||\dot{x}_1(t) - \dot{x}_0(t)|| = ||v_0(t) - \dot{x}_0(t)|| \leq \rho(t) \quad (1.13)$$

a.e. $t \in I$ and

$$\begin{aligned}
||x_1(t) - x_0(t)|| &\leq ||x_0 - \gamma(t_0)|| + |\int_{t_0}^{t} ||\dot{x}_1(s) - \dot{x}_0(s)||\,ds| \\
&\leq r + |\int_{t_0}^{t} \rho(s)\,ds| .
\end{aligned} \quad (1.14)$$

Note in particular that $r + |\int_{t_0}^{t} \rho(s)\,ds| \leq L(t)$, so that for $t \in J$ we have

$$||x_1(t) - x_0(t)|| \leq b . \quad (1.15)$$

By repeating the same argument as above, we can now take $v_1(t)$ in such a way that

$$v_1(t) \in F(t, x_1(t)) \quad \text{and} \quad ||v_1(t) - \dot{x}_1(t)|| = \text{dist}\,(\dot{x}_1(t), F(t, x_1(t))) \tag{1.16}$$

a.e. $t \in I$. On the other hand, taking into account (1.15), from (1.16), and the definition of the Hausdorff distance, we deduce

$$||v_1(t) - \dot{x}_1(t)|| \leq h(F(t, x_0(t)), F(t, x_1(t))) \leq l(t)||x_0(t) - x_1(t)|| \tag{1.17}$$

and this in turn implies that $v_1(t)$ is integrable on J. Thus, we can iterate the construction, defining

$$x_2(t) = x_0 + \int_{t_0}^{t} v_1(s)\,ds \tag{1.18}$$

for $t \in J$. As for $x_1(t)$, we need some estimations concerning $x_2(t)$ and its derivative. According to (1.17) and (1.14), we infer

$$\begin{aligned}
||\dot{x}_2(t) - \dot{x}_1(t)|| &= ||v_1(t) - \dot{x}_1(t)|| \\
&\leq l(t)||x_0(t) - x_1(t)|| \\
&\leq l(t)\left[r + \left|\int_{t_0}^{t} \rho(s)\,ds\right|\right]
\end{aligned} \tag{1.19}$$

a.e. $t \in J$. Moreover, by (1.19) and (1.13),

$$\begin{aligned}
||\dot{x}_2(t) - \dot{x}_0(t)|| &\leq ||\dot{x}_2(t) - \dot{x}_1(t)|| + ||\dot{x}_1(t) - \dot{x}_0(t)|| \\
&\leq l(t)\left[r + \left|\int_{t_0}^{t} \rho(s)\,ds\right|\right] + \rho(t)
\end{aligned} \tag{1.20}$$

a.e. $t \in J$. We also have, by virtue of (1.19),

$$\begin{aligned}
||x_2(t) - x_1(t)|| &\leq \left|\int_{t_0}^{t} ||\dot{x}_2(s) - \dot{x}_1(s)||\,ds\right| \\
&\leq \left|\int_{t_0}^{t} l(s)\left[r + \left|\int_{t_0}^{s} \rho(\sigma)\,d\sigma\right|\right]ds\right| \\
&\leq rm(t) + \left|\int_{t_0}^{t} \rho(s)[m(t) - m(s)]\,ds\right|
\end{aligned} \tag{1.21}$$

for each $t \in J$. Moreover, using (1.14),

$$||x_2(t) - x_0(t)|| \leq ||x_2(t) - x_1(t)|| + ||x_1(t) - x_0(t)||$$
$$\leq r(1 + m(t))$$
$$+ \left| \int_{t_0}^{t} \rho(s)[1 + (m(t) - m(s))] \, ds \right| \quad (1.22)$$

for each $t \in J$. This last expression is less than or equal to $L(t)$, so that

$$||x_2(t) - x_0(t)|| \leq L(t) \leq b \quad (1.23)$$

for $t \in J$. Take now $v_2(t)$ in such a way that

$$v_2(t) \in F(t, x_2(t)) \quad \text{and} \quad ||v_2(t) - \dot{x}_2(t)|| = \text{dist}\,(\dot{x}_2(t), F(t, x_2(t))) \quad (1.24)$$

a.e. $t \in J$. By repeating again the argument, we get

$$||v_2(t) - \dot{x}_2(t)|| \leq h(F(t, x_1(t)), F(t, x_2(t))) \leq l(t) ||x_1(t) - x_2(t)|| \quad (1.25)$$

a.e. $t \in J$, which allows us to conclude that $v_2(t)$ is integrable. Therefore, the procedure can be further iterated, and it can be formalized by induction. We finally found that for each $i = 0, 1, 2, \ldots$,

$$x_{i+1}(t) = x_0 + \int_{t_0}^{t} v_i(s) \, ds \quad (1.26)$$

for $t \in J$,

$$v_i(t) \in F(t, x_i(t)) \quad \text{and} \quad ||v_i(t) - \dot{x}_i(t)|| = \text{dist}\,(\dot{x}_i(t), F(t, x_i(t))) \quad (1.27)$$

a.e. $t \in J$,

$$||\dot{x}_{i+1}(t) - \dot{x}_i(t)|| \leq l(t) \left[r \frac{(m(t))^{i-1}}{(i-1)!} + \left| \int_{t_0}^{t} \rho(s) \frac{(m(t) - m(s))^{i-1}}{(i-1)!} \, ds \right| \right] \quad (1.28)$$

for $i \geq 1$ and a.e. $t \in J$,

$$||x_{i+1}(t) - x_i(t)|| \leq r \frac{(m(t))^i}{i!} + \left| \int_{t_0}^{t} \rho(s) \frac{(m(t) - m(s))^i}{i!} \, ds \right| \quad (1.29)$$

for $i \geq 0$ and $t \in J$. Finally, using (1.29), (1.28) and the inequality $1 + z + \ldots + z^j/j! \leq e^z$ ($z \geq 0$), we obtain for each $j = 1, 2, \ldots$ and $t \in J$

$$||x_j(t) - x_0(t)|| \leq L(t) \leq b \tag{1.30}$$

and

$$||\dot{x}_j(t) - \dot{x}_0(t)|| \leq l(t)L(t) + \rho(t) \tag{1.31}$$

for a.e. $t \in J$. We are now ready to get the conclusion. We see from (1.29) that the sequences $\{x_i(t)\}$ converges (uniformly) on J and from (1.28) that $\{v_i(t)\} = \{\dot{x}_{i+1}(t)\}$ converges a.e. on J. Let $x(t)$ and $v(t)$ be the respective limits. Inequality (1.31) enables us to apply the Lebesgue dominated convergence Theorem to (1.26). So we obtain

$$x(t) = \lim_j x_{j+1}(t) = x_0 + \lim_j \int_{t_0}^t v_j(s)\, ds = x_0 + \int_{t_0}^t v(s)\, ds$$

and we get the first conclusion that $x(t)$ is absolutely continuous with $\dot{x}(t) = v(t)$ a.e. on J. On the other hand, using continuity of $F(t,x)$ with respect to x, from (1.27) we get

$$v(t) \in F(t, x(t))$$

a.e. $t \in J$. Hence, $x(t)$ is a solution. Inequalities (1.9) and (1.10) easily follow from (1.30) and (1.31). ∎

1.3 Appendix

We give here the proofs of Theorems 1.2 and 1.3. Recall that a function is said to be *simple* if the set of its values is finite. If C is a convex set, then by ext C we denote the set of its extreme points.

We start by the following lemma.

Lemma 1.1 *Let $g(t,x)$ be a simple Borel map from $[0,1)^{n+1}$ to $\mathbb{R}^n$. Then, $G(t,x) = \mathbf{K}_x g(t,x)$, as a set valued map from $[0,1)^{n+1}$ to $\mathbb{R}^n$, is a Borel map.*

Proof. Let us denote by $\mathcal{A}$ the σ-algebra of Borel sets in $[0,1)^{n+1}$. Let $\{a_i, 1 \leq i \leq l\}$ with $l \geq 1$ be the set of values of g. There exist pairwise disjoint sets $B_i \in \mathcal{A}$ $(i = 1, \ldots, l)$ such that $B_i \subseteq [0,1)^{n+1}$, $\cup B_i = [0,1)^{n+1}$, and

$$g(t,x) = \sum_{i=1}^{l} a_i \chi_{B_i}(t,x) \ .$$

By virtue of (1.4), it follows that for each pair $(t,x) \in [0,1)^{n+1}$ there exists $I \subseteq \{1,\ldots,l\}$ ($I \neq \emptyset$) such that

$$G(t,x) = \overline{\text{co}}\{a_i, i \in I\} \ .$$

On the other hand, let $(t_0, x_0) \in [0,1)^{n+1}$. Let us fix $I \subseteq \{1,\ldots,l\}$ in such a way that $\{a_i, i \in I\} = \text{ext}\,\overline{\text{co}}\{a_i, i \in I\}$. It is clear that $G(t_0, x_0) = \overline{\text{co}}\{a_i, i \in I\}$ if and only if:

1) $\forall i \in I$, $\forall k \in \mathbb{N}^*$, one has $\mu(\{x \in \mathbb{R}^n : x \in B_{1/k}(x_0) \text{ and } (t_0, x) \in B_i\}) > 0$

2) $\forall j \notin I$, with $a_j \notin \overline{\text{co}}\{a_i, i \in I\}$, $\exists k \in \mathbb{N}^*$ such that $\mu(\{x \in \mathbb{R}^n : x \in B_{1/k}(x_0) \text{ and } (t_0, x) \in B_j\}) = 0$.

Next we want to prove the following:

Claim $\{(t_0, x_0) \in [0,1)^{n+1} : G(t_0, x_0) = \overline{\text{co}}\{a_i, i \in I\}\} \in \mathcal{A}$, where $I \subseteq \{1,\ldots,l\}$ has been fixed as above.

For each i and k, we define the map

$$\psi_{i,k} : (t_0, x_0) \mapsto \mu(\{x \in \mathbb{R}^n : x \in B_{1/k}(x_0) \text{ and } (t_0, x) \in B_i\}) \ .$$

We have

$$\psi_{i,k}(t_0, x_0) = \int_{\mathbb{R}^n} \chi_{B_{1/k}(x_0)}(x) \cdot \chi_{B_i}(t_0, x)\, dx$$

$$= \int_{\mathbb{R}^n} \chi_{\{(x,x_0) : ||x - x_0|| < \frac{1}{k}\}}(x, x_0) \cdot \chi_{B_i}(t_0, x)\, dx \ .$$

The map $(x, x_0, t_0) \mapsto \chi_{\{(x,x_0) : ||x - x_0|| < \frac{1}{k}\}}(x, x_0) \cdot \chi_{B_i}(t_0, x)$ is a Borel map, hence by Tonelli's Theorem ([55], p. 85), $\psi_{i,k}(t_0, x_0)$ is a Borel map, as well. Finally, we note that

$$\{(t_0, x_0) \in [0,1)^{n+1} : G(t_0, x_0) = \overline{\text{co}}\{a_i, i \in I\}\}$$
$$= \left(\cap_{i \in I} \cap_{k \in \mathbb{N}^*} (\psi_{i,k})^{-1}((0, +\infty))\right)$$
$$\cap \left(\cap_{\{j : a_j \notin \overline{\text{co}}\{a_i : i \in I\}\}} \cup_{k \in \mathbb{N}^*} (\psi_{j,k})^{-1}(\{0\})\right) \ .$$

The claim easily follows. We are now able to achieve the proof of the Lemma. Given any open set $\Omega \subseteq \mathbb{R}^n$, we observe that

$$G^{-1}(\Omega) = \{(t,x) \in [0,1)^{n+1} : G(t,x) \cap \Omega \neq \emptyset\}$$
$$= \bigcup_I \{(t,x) \in [0,1)^{n+1} : G(t,x) = \overline{co}\{a_i, i \in I\}\}$$

where the union is taken over all the sets $I \subseteq \{1,\ldots,l\}$, $I \neq \emptyset$, such that $\overline{co}\{a_i, i \in I\} \cap \Omega \neq \emptyset$ and $\{a_i, i \in I\} = \text{ext}\,\overline{co}\{a_i, i \in I\}$.

In other words, $G^{-1}(\Omega)$ is a finite union of sets which belong to the σ-algebra $\mathcal{A}$. Hence, $G^{-1}(\Omega) \in \mathcal{A}$. The Lemma is proved. ■

Proof of Theorem 1.2. Assume first that $f(t,x)$ is a Borel map. Then, it is sufficient to prove that for each open set $\Omega \subseteq \mathbb{R}^n$ and each $m = (m_0, m_1, \ldots, m_n) \in \mathbb{Z}^{n+1}$ we have that

$$\left(F|_{m+[0,1)^{n+1}}\right)^{-1}(\Omega)$$

is a Borel set. Indeed,

$$F^{-1}(\Omega) = \bigcup_{m \in \mathbb{Z}^{n+1}} \{(t,x) \in m+[0,1)^{n+1} : F(t,x) \cap \Omega \neq \emptyset\}$$
$$= \bigcup_{m \in \mathbb{Z}^{n+1}} \left(F|_{m+[0,1)^{n+1}}\right)^{-1}(\Omega).$$

For simplicity we consider only the case $m = 0$, but the same arguments apply for each $m \in \mathbb{Z}^{n+1}$. Let $\{f_p\}$ be a sequence of simple, Borel maps from $[0,1)^{n+1}$ to $\mathbb{R}^n$ such that

$$\lim_{p \to \infty} f_p(t,x) = f(t,x) \tag{1.32}$$

uniformly on $[0,1)^{n+1}$. Such a sequence exists since f is bounded on $[0,1)^{n+1}$. Set $F_p(t,x) = \mathbf{K}_x f_p(t,x)$. According to Lemma 1.1, $\{F_p(t,x)\}$ is a sequence of Borel set valued maps from $[0,1)^{n+1}$ to $\mathbb{R}^n$. For each $(t,x) \in [0,1)^{n+1}$ and each p, we have

$$h(F_p(t,x), F(t,x)) \leq \|f_p - f\|_{L^\infty([0,1)^{n+1})}. \tag{1.33}$$

For any $y \in \mathbb{R}^n$, let us consider now the maps $\gamma_p(t,x) = \text{dist}\,(y, F_p(t,x))$ and $\gamma(t,x) = \text{dist}\,(y, F|_{[0,1)^{n+1}}(t,x))$. From (1.32) and (1.33) it follows that γ_p converges to γ, as $p \to \infty$.

By Castaing's Characterization Theorem (see [8], Theorem 8.3.1 p. 319), for each $y \in \mathbb{R}^n$ and each p, γ_p is a Borel map. Hence, for each $y \in \mathbb{R}^n$, γ

is a Borel map. Using again Castaing's Characterization Theorem, we finally conclude that also $F|_{[0,1)^{n+1}}$ is a Borel map.

The argument when f is only (Lebesgue) measurable is given in the proof of Theorem 1.3. ∎

Proof of Theorem 1.3. Since $f = f(t, x)$ is locally bounded and measurable, there exists a locally bounded Borel function $\tilde{f} = \tilde{f}(t, x)$ such that

$$\tilde{f}(t, x) = f(t, x)$$

for a.e. $(t, x) \in \mathbb{R}^+ \times \mathbb{R}^n$ (see [122], p. 56). Let $N_1 = \{(t, x) : \tilde{f}(t, x) \neq f(t, x)\}$. Then there exists a Borel set $\tilde{N}_1$ such that $N_1 \subseteq \tilde{N}_1$ and $\mu(\tilde{N}_1) = 0$. By Tonelli's Theorem,

$$0 = \int\int_{\mathbb{R}^+ \times \mathbb{R}^n} \chi_{\tilde{N}_1}(t, x) \, dt dx = \int_0^\infty \left(\int_{\mathbb{R}^n} \chi_{\tilde{N}_1}(t, x) \, dx \right) dt \ .$$

Note that the map $t \mapsto \int_{\mathbb{R}^n} \chi_{\tilde{N}_1} \, dx$ is non-negative. Hence, the Borel set $N_2 = \{t \in [0, +\infty) : \int_{\mathbb{R}^n} \chi_{\tilde{N}_1} \, dx > 0\}$ is of measure zero.

We claim that

$$\forall t \notin N_2, \ \forall x \in \mathbb{R}^n \quad \tilde{F}(t, x) = F(t, x) \tag{1.34}$$

where $\tilde{F}(t, x) = \mathbf{K}_x \tilde{f}(t, x)$. Indeed, if $t \notin N_2$, then $\int_{\mathbb{R}^n} \chi_{\tilde{N}_1} \, dx = 0$, that is $\mu(\{x : (t, x) \in \tilde{N}_1\}) = 0$. This means that $\tilde{f}(t, x) = f(t, x)$ for a.e. $x \in \mathbb{R}^n$, and this in turn implies (1.34) for every $x \in \mathbb{R}^n$.

As a first consequence, we deduce from (1.34) that Filippov solutions of the systems

$$\dot{x} = f(t, x) \quad \text{and} \quad \dot{x} = \tilde{f}(t, x)$$

agree. Moreover, according to Proposition 1.1 and Theorem 1.2, the set valued map $\tilde{F}(t, x)$ is everywhere defined, with nonempty convex, compact values, and Borel measurable. As already noticed, for fixed $x \in \mathbb{R}^n$, the map $t \mapsto \tilde{F}(t, x)$ can be reviewed as the composition of $\tilde{F} : \mathbb{R}^+ \times \mathbb{R}^n \to \mathbb{R}^n$ and the Borel map $t \mapsto (t, x) : \mathbb{R}^+ \to \mathbb{R}^+ \times \mathbb{R}^n$. Hence, it is a Borel map, as well. But for every open set $\Omega \subseteq \mathbb{R}^n$,

$$\begin{aligned} F(\cdot, x)^{-1}(\Omega) &= \{t \in \mathbb{R}^+ \setminus N_2 : F(t, x) \cap \Omega \neq \emptyset\} \cup N_3 \\ &= \{t \in \mathbb{R}^+ \setminus N_2 : \tilde{F}(t, x) \cap \Omega \neq \emptyset\} \cup N_3 \end{aligned}$$

where N_3 is some set of measure zero, $N_3 \subseteq N_2$.

This implies that the map $t \mapsto F(t,x)$ is (Lebesgue) measurable for each $x \in \mathbb{R}^n$, as required. In the same way,

$$F^{-1}(\Omega) = \{(t,x) \in (\mathbb{R}^+ \setminus N_2) \times \mathbb{R}^n : \tilde{F}(t,x) \cap \Omega \neq \emptyset\} \cup N_4$$

where $N_4 \subset N_2 \times \mathbb{R}^n$ is some set of measure 0. It follows that F is measurable, which completes the proof of Theorem 1.2. ∎

Chapter 2

Time invariant systems

In this chapter we use the same notation already introduced at the beginning of Chapter 1. Moreover, for each Lebesgue measurable, essentially bounded function $u : [0, +\infty) \to \mathbb{R}^m$, we denote the L_∞ norm of $u(\cdot)$ by

$$\|u(\cdot)\|_\infty = \operatorname*{ess\,sup}_{t \geq 0} \|u(t)\| < +\infty \;.$$

As explained in the Introduction, studies on stability and stabilizability have been developed first in the framework of linear theory and subsequently, during the last two decades, for nonlinear control systems which are time invariant and sufficiently "smooth". In order to enlighten the most important concepts involved and to survey the main results obtained so far, in the present chapter our exposition will be therefore limited to such a restricted class of systems. In fact, it seems convenient to start by a short digression about linear systems: since in this context technical difficulties are highly reduced, we can take advantage of a more immediate appeal to intuition.

2.1 The linear case

Linear systems have been widely studied for a long time and a rather complete picture is today available. It is natural to take it as a reference model for possible nonlinear developments.

2.1.1 Stability

In particular, the relationship between internal and external stability is well understood in the case of the finite-dimensional, time invariant linear system

$$\dot{x} = Ax + Bu \qquad (2.1)$$

(here, A and B are real matrices of appropriate dimensions). The unforced associated system has the form

$$\dot{x} = Ax \ . \qquad (2.2)$$

It is well known that the asymptotic behavior of its solutions depends on the eigenvalues of A. More precisely, A is *Hurwitz* (i.e., all its eigenvalues lie on the open left half complex plane) if and only if all the solutions converge to the origin for $t \to +\infty$. Moreover, A is *stable* (i.e., all its eigenvalues lie on the closed left half complex plane and the possible eigenvalues on the imaginary axis are simple[1]) if and only if all the solutions are bounded and the solutions issuing from a sufficient small neighborhood of the origin remain near the origin for all $t \geq 0$.

If A is Hurwitz and if $u : [0, +\infty) \to \mathbb{R}^m$ is an admissible input (namely, a Lebesgue measurable function such that $\|u(\cdot)\|_\infty < +\infty$), then the variation of constants formula can be used to prove that there exist positive numbers $\alpha, \gamma_1, \gamma_2$ such that

$$\|x(t; x_0, u(\cdot))\| \leq \gamma_1 \|x_0\| e^{-\alpha t} + \gamma_2 \|u(\cdot)\|_\infty \qquad (2.3)$$

for each $x_0 \in \mathbb{R}^n$ and $t \geq 0$. Conversely, if (2.3) holds for each initial state, each admissible input and each $t \geq 0$, then the special choice $u \equiv 0$ shows that A must be Hurwitz.

Inequality (2.3) admits the following interpretation: for t large enough, the effect of the initial conditions is negligible, and the solutions are ultimately bounded by a term which is related to the input energy (measured by its L_∞ norm) by means of the constant "gain" γ_2. This reflects the distinction between transient and steady state in the classical engineering literature.

Beside (2.3), we are also interested in the following condition: there exist some constants $\gamma_1, \gamma_2 > 0$ for which

$$\|x(t; x_0, u(\cdot))\| \leq \gamma_1 \|x_0\| + \gamma_2 \|u(\cdot)\|_\infty \qquad (2.4)$$

for each $t \geq 0$, each x_0 and each admissible input $u(\cdot)$.

Of course, (2.4) is implied by (2.3) and in turn, (2.4) implies that A is stable.

---

[1] This means that their algebraic and geometric multiplicities coincide.

Inequality (2.4) can be interpreted by saying that when the input is bounded, then the solutions are bounded by a term which depends linearly on the energy initially stored in the system (measured by the norm of the initial state) and the energy due to the input supply.

Inequalities (2.3) and (2.4) are appropriate tools in order to describe the effect of inputs and initial conditions on the evolution of the linear system defined by (2.1). Sometimes, we will refer to (2.3) [respectively, (2.4)] by saying that (2.1) has the *strong* [respectively, *weak*] *finite gain property*[2].

To resume the previous discussion, we can single out in particular the following conclusions.

Proposition 2.1 *If A is Hurwitz, then (2.1) has the (strong, and hence also the) weak finite gain property.*

Proposition 2.2 *If (2.1) has the weak finite gain property, then A is a stable matrix.*

The meaning of Proposition 2.1 is captured by the following informal expression: *internal stability implies external stability*. The converse is true only under additional assumptions.

Recall that (2.1) is said to be *completely controllable* when it happens that any initial state can be steered to any desired final state in finite time by a suitable choice of an open loop control $u = u(t)$. It is well known that (2.1) is completely controllable if and only if the rank of the matrix $\begin{bmatrix} B \vdots AB \vdots \ldots \vdots A^{n-1}B \end{bmatrix}$ is maximal (i.e., it is equal to n).

Proposition 2.3 *Let (2.1) be completely controllable. If it has the weak finite gain property, then A is Hurwitz.*

Proof. It is not difficult to prove that if a linear system has the weak finite gain property then the so called *impulse response*

$$\int_0^t \|e^{sA}B\|\,ds$$

is bounded for $t \geq 0$. Arguing as in [133] p. 257, this in turn implies that

---

[2] Usually, a complete model of a linear system includes an observation map of the form $y = Cx$. The variable y is called the *output*. In this context, the finite gain property should be referred to outputs, instead of state evolution, and it is also called *bounded input bounded output* (in short, BIBO) stability.

$$\lim_{t \to +\infty} u(t) = 0 \quad \Longrightarrow \quad \lim_{t \to +\infty} x(t; 0, u(\cdot)) = 0 \ .$$

Pick now an arbitrary $x_0 \in \mathbb{R}^n$. The complete controllability assumption implies that there is an input $u_1(t)$ and some $T > 0$ such that $x_0 = x(T; 0, u_1(\cdot))$. Define the new input

$$u_2(t) = \begin{cases} u_1(t) & \text{for } t \in [0, T] \\ 0 & \text{for } t \geq T \ . \end{cases}$$

Clearly, $\lim_{t \to +\infty} x(t; 0, u_2(\cdot)) = 0$. But for $t \geq T$ the input vanishes, so that the motion depends only on the action of the unforced system. It remains to note that $x(T; 0, u_2(\cdot)) = x(T; 0, u_1(\cdot)) = x_0$. ∎

The simple example

$$\begin{cases} \dot{x} = ax + u \\ \dot{y} = 0 \end{cases} \tag{2.5}$$

with $a = -1$ shows that the controllability assumption cannot be dropped out, in general.

Also the converse of Proposition 2.2 is false, but as we shall see at the end of Section 2.1.3, we can obtain a positive result making use of linear feedback.

2.1.2 Internal stabilization

A *linear feedback* is a feedback law of the form $u = Fx$ (where F is a matrix of appropriate dimensions). The effect of applying a linear feedback to (2.1) consists in replacing the matrix A by the modified matrix $A + BF$. Hence, if A is not Hurwitz, it makes sense to ask whether there exists a linear feedback $u = Fx$ such that all the trajectories of the associated closed loop system converge to zero (i.e., the matrix $A + BF$ is Hurwitz). If this happens we also say that (2.1) is *internally stabilizable*, and the linear map $u = Fx$ is called a *stabilizer*.

Notice that, according to Proposition 2.1, if the system is internally stabilizable, then the closed loop system has the property (2.4) (in fact, it has the strong finite gain property). We can also say that *internal stabilization implies external stabilization*.

A variety of necessary and sufficient conditions for internal stabilizability of a linear system can be found in the literature. For the moment, we limit ourselves to recall the following one (the proof can be found in any common textbook about linear systems, for instance [156]).

Proposition 2.4 *Let $\mathcal{V} \subseteq \mathbb{R}^n$ be the linear space engendered by all the eigenvectors (in the usual and generalized sense) corresponding to eigenvalues of A with non-negative real part. System (2.1) is internally stabilizable if and only if*

$$\mathcal{V} \subseteq [B \vdots AB \vdots \ldots \vdots A^{n-1}B] .$$

In particular, we have that complete controllability implies internal stabilizability.

2.1.3 External stabilization

Example (2.5) with $a = 1$ shows that the weak finite gain property can be achieved by means of feedback even if the system is not necessarily internally stabilizable. The aim of this section is to find out a necessary and sufficient condition for the existence of externally stabilizing (in the sense of the finite gain property) feedback laws. First of all, we remark that the weak finite gain property is invariant under linear changes of coordinates in the state space. Thus, it is not restrictive to assume that (2.1) is in Kalman canonical form, namely

$$\begin{cases} \dot{x}_1 = A_{11}x_1 + A_{12}x_2 + B_1 u \\ \dot{x}_2 = A_{22}x_2 \end{cases} \quad (2.6)$$

where $x = (x_1, x_2) \in \mathbb{R}^q \times \mathbb{R}^{n-q}$ $(0 \leq q \leq n)$ and the pair (A_{11}, B_1) define a q-dimensional completely controllable system

$$\dot{x}_1 = A_{11}x_1 + B_1 u . \quad (2.7)$$

Our argument is based on the following lemma.

Lemma 2.1 *System (2.1) has the weak finite gain property if and only if A_{11} is Hurwitz and A_{22} is stable.*

Proof. Assume that A_{11} is Hurwitz and A_{22} stable. Then, in particular, $x_2(t)$ is bounded for each initial state. Hence, for each admissible input $u(\cdot)$ we can interpret the subsystem

$$\dot{x}_1 = A_{11}x_1 + A_{12}x_2 + B_1 u \quad (2.8)$$

as having a bounded input $v(t) = A_{12}x_2(t) + B_1 u(t)$. According to Proposition 2.1, since A_{11} is Hurwitz, $x_1(t)$ is bounded for each initial state, as well.

As far as the converse is concerned, we already noticed that if (2.4) holds then A must be stable and of course the same is true for A_{22}. To prove that A_{11} is Hurwitz, we observe that if the overall system has the weak finite gain property, then the q-dimensional completely controllable system (2.7) has the weak finite gain property, as well (this can be seen by considering solutions of (2.6) which correspond to initial conditions of the form $(\bar{x}_1, 0)$). The statement is therefore a consequence of Proposition 2.3. ∎

It follows immediately from Lemma 2.1 that:

Proposition 2.5 *Let the linear system (2.1) be given. There exists a linear feedback $u = Fx$ such that the closed loop system satisfies (2.4) if and only if A_{22} is stable.*

Proposition 2.5 implies in particular that whenever A is stable, then it is possible to recover the weak finite gain property by means of linear feedback.

2.1.4 Quadratic forms

Internal stability analysis of linear systems can be performed also by means of certain auxiliary quadratic functions i.e., functions of the form $V(x) = x^t P x$. Indeed, it is possible to prove that A is Hurwitz [stable] if and only if there exists a symmetric, positive definite matrix P such that

$$\dot{V}(x) \underset{\text{def}}{=} 2x^t P A x = x^t (PA + A^t P) x$$

is negative definite [negative semi-definite].

Quadratic forms can be used to characterize internal stabilizability as well. We recall the following result, which is of some interest for nonlinear developments.

Proposition 2.6 *The linear system (2.1) is internally stabilizable if and only if there exists a real, positive definite, symmetric matrix P such that*

$$\{x \in \mathbb{R}^n : x^t P A x \geq 0\} \cap \ker B^t P = \{0\} . \tag{2.9}$$

Moreover, if (2.9) holds then there exists $\alpha_0 > 0$ such that the stabilizing feedback can be taken of the form

$$u = -\alpha B^t P x \tag{2.10}$$

for any $\alpha \geq \alpha_0$.

Proof. The "only if" part is easy. If the system admits a linear stabilizer $u = Fx$, then the matrix $A + BF$ of the closed loop system is Hurwitz. Hence, there is a quadratic function $V(x) = x^t P x$ such that

$$\dot{V}(x) = 2(x^t P A x + x^t P B F x) < 0$$

for each $x \neq 0$. When $x \in \ker B^t P$ this clearly reduces to $x^t P A x < 0$.

We prove now the "if" part. Define the feedback law as indicated in (2.10) and let $V(x) = x^t P x$. With respect to the closed loop system, we have

$$\dot{V}(x) = 2x^t P(A + BF)x = 2x^t P(A - \alpha B B^t P)x = 2x^t P A x - 2\alpha \|B^t P x\|^2 .$$

This last expression is homogeneous. In order to study its sign, we can limit ourselves to the set $S = \{x \in \mathbb{R}^n : \|x\| = 1\}$.

If $x \in S \cap \ker B^t P$, then $\dot{V}(x) < 0$ because of (2.9). Since $\dot{V}(x)$ is continuous, there exists a relatively open set $U \subset S$ such that $S \cap \ker B^t P \subset U$ and $\dot{V}(x) < 0$ for each $x \in U$.

The complement of U in S is compact. Hence, $\min_{S \setminus U} \|B^t P x\|^2 = m > 0$. Let $M = \max_S x^t P A x$. It is clear that if α is larger than M/m, then $\dot{V}(x) < 0$ also for $x \in S \setminus U$. The proof is complete. ∎

For an internally stabilizable system, the matrix P satisfying condition (2.9) is not unique. In fact, we have a more precise result (see [42]): if (2.1) is internally stabilizable, then there exists a matrix P for which (2.9) holds, and the feedback law (2.10) works with $\alpha = \frac{1}{2}$. For such a matrix P, negative definiteness of $\dot{V}(x)$ is clearly equivalent to

$$PA + A^t P - PBB^t P = -Q \tag{2.11}$$

for some symmetric positive definite matrix Q.

Vice-versa, for fixed Q, (2.11) can be interpreted as a matrix equation in the unknown P. If a positive definite solution $P = P(Q)$ of (2.11) exists, then it is unique and satisfies (2.9). Moreover, it can be used to define a stabilizing feedback of the form (2.10) with $\alpha = \frac{1}{2}$. This conclusion does not depend on the choice of Q, so that it is usual to take $Q = I$, the identity matrix.

Equation (2.11) is a form of the so-called *algebraic Riccati matrix equation*.

Unfortunately, the characterization of external stability by means of auxiliary quadratic functions is not as much plain. We already know that (2.3) holds if and only if A is Hurwitz. It is not difficult to prove the following additional characterization.

26   Time invariant systems

Proposition 2.7 *Inequality (2.3) holds if and only if there exists a positive definite, symmetric real matrix P which enjoys the following property: for each $R > 0$ there exists $N > 0$ such that*

$$x^t P(Ax + Bu) < 0 \qquad (2.12)$$

for each $x \in \mathbb{R}^n$, $u \in \mathbb{R}^m$, subject to the conditions $\|u\| \leq R$ and $\|x\| \geq N$.

At this point, it could be tempting to conjecture that a necessary and sufficient condition for the weak finite gain property is obtained in a similar way, by simply substituting the strict inequality in (2.12) by a weak one. As we shall see later, in this way one obtains a condition which is actually sufficient, but not necessary, as shown by the extremely simple example (2.5). Indeed, we already know that (2.5) with $a = -1$ has the weak finite gain property. Setting for instance $u = 1$ and given a matrix $P = \begin{pmatrix} p_{11} & p_{12} \\ p_{12} & p_{22} \end{pmatrix}$, the expression in (2.12) is easily computed:

$$(p_{11}x_1 + p_{12}x_2)(1 - x_1) \ .$$

As Figure 2.1 shows, this expression takes both positive and negative values for $\|x\|$ arbitrarily large, and for any choice of p_{11}, p_{12}.

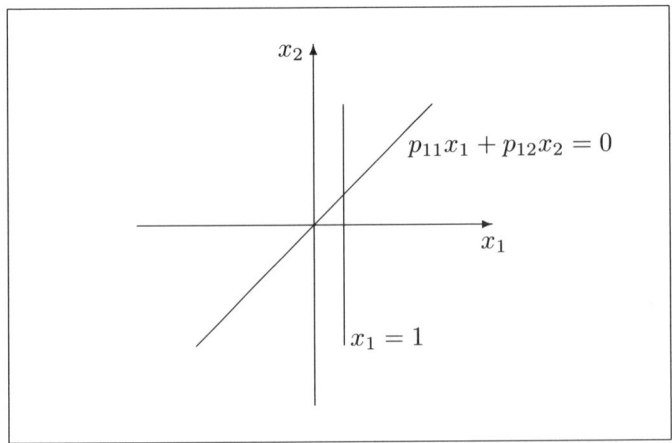

Figure 2.1: Regions where the expression $(p_{11}x_1 + p_{12}x_2)(1 - x_1)$ changes sign

2.2 Nonlinear systems: stability

From now on, we turn our attention to nonlinear systems. In particular, this section is devoted to internal and external notions of stability, Liapunov functions and the related theorems.

2.2.1 Internal notions

The internal notions of stability for the input system (1) are nothing else that the classical notions of stability for the unforced associated system (2). For the moment, we can therefore limit ourselves to consider systems of ordinary differential equations of the form

$$\dot{x} = f(x), \qquad x \in \mathbb{R}^n . \tag{2.13}$$

We assume that f is a continuous vector field. This guarantees that for each initial state x_0, there exists at least one (classical) solution $x(t)$ such that $x(0) = x_0$. Moreover, every Carathéodory solution is actually a classical one, so that throughout this chapter the symbol $\mathcal{S}_{x_0}$ will denote the set of all the classical solutions.

Definition 2.1 *We say that (2.13) is* (Liapunov) stable *at the origin (or that the origin is stable for (2.13)) if for each $\varepsilon > 0$ there exists $\delta > 0$ such that for each x_0 with $||x_0|| < \delta$ and all the solutions $x(\cdot) \in \mathcal{S}_{x_0}$ the following holds: $x(\cdot)$ is right continuable for $t \in [0, +\infty)$ and*

$$||x(t)|| < \varepsilon \quad \forall t \geq 0 .$$

Note that if the origin is stable, then it is an equilibrium position for (2.13) i.e., $f(0) = 0$.

Definition 2.2 *We say that (2.13) is* Lagrange stable *(or that it has the property of uniform boundedness of solutions) if for each $R > 0$ there exists $S > 0$ such that for $||x_0|| < R$ and all the solutions $x(\cdot) \in \mathcal{S}_{x_0}$ one has that $x(\cdot)$ is right continuable for $t \in [0, +\infty)$ and*

$$||x(t)|| < S , \quad \forall t \geq 0 .$$

In the linear case Liapunov stability and Lagrange stability imply each other; in general, it should be clear that they are distinct properties.

Definition 2.3 *We say that system (2.13) is* locally asymptotically stable at the origin *(or that the origin is locally asymptotically stable for (2.13)) if it is stable at the origin and, in addition, the following condition holds: there exists $\delta_0 > 0$ such that*

$$\lim_{t \to +\infty} \|x(t)\| = 0$$

for each x_0 such that $\|x_0\| < \delta_0$, and all the solutions $x(\cdot) \in S_{x_0}$.

The origin is said to be globally asymptotically stable *if δ_0 can be taken as large as desired.*

Remark 2.1 When dealing with systems without uniqueness, one should distinguish between weak and strong notions. The previous definitions are *strong* notions in the sense that the properties are required to hold for all the solutions, and not only for some of them.

Remark 2.2 Definitions 2.1 and 2.3 can be referred to any equilibrium position, that is any point x_0 such that $f(x_0) = 0$. The choice $x_0 = 0$ implies no loss of generality. ■

Speaking about systems with inputs, it may be convenient to adopt a modified terminology.

Definition 2.4 *Consider an input system of the form (1). We say that it is* internally (Liapunov) stable at the origin *[respectively,* internally Lagrange stable, internally locally *or* globally asymptotically stable at the origin*] whenever the unforced associated system (2) is stable at the origin [respectively, Lagrange stable, locally or globally asymptotically stable at the origin].*

When the system is not linear, and we do not have information about the structure of the right hand side of (2.13), any attempt to derive stability criteria of algebraic nature is hopeless. On the contrary, the method of auxiliary functions can be generalized. Of course, we cannot expect to succeed in stability analysis of nonlinear systems by means of auxiliary functions which are actually "quadratic" or polynomial functions (these are obviously non-essential details, linked to the linearity context of the previous section). Rather, we can try to figure out a more abstract notion which retains the main features of quadratic functions. This leads to the idea of Liapunov function.

Actually, for each concept of stability there is a corresponding concept of Liapunov function[3]. Recall the notation $B_r = \{x \in \mathbb{R}^n : \|x\| < r\}$ and $B^r = \{x \in \mathbb{R}^n : \|x\| > r\}$.

Definition 2.5 *A smooth weak Liapunov function in the small is a real map $V(x)$ which is defined on B_r for some $r > 0$, and fulfills the following properties:*
(i) $V(0) = 0$
(ii) $V(x) > 0$ for $x \neq 0$
(iii) $V(x)$ is of class C^1 on B_r
(iv) $\nabla V(x) \cdot f(x) \leq 0$ for each $x \in B_r$.

When a real function $V(x)$ satisfies (ii), it is usual to say that it is *positive definite*. The function

$$\dot{V}(x) \underset{\text{def}}{=} \nabla V(x) \cdot f(x)$$

is called the *derivative of V with respect to (2.13)*. Condition (iv) means that $\dot{V}$ is *semi-definite negative*.

A real function $V(x)$ is said to be *radially unbounded* if it is defined on B^r for some $r > 0$, and

$$\lim_{\|x\| \to +\infty} V(x) = +\infty \ .$$

This is equivalent to say that the level sets $\{x \in \mathbb{R}^n : V(x) \leq a\}$ are bounded for each $a \in \mathbb{R}$.

Definition 2.6 *A function $V(x)$ defined on B^r for some $r > 0$, which is radially unbounded and fulfills (iii) and (iv) of Definition 2.5 (with B_r replaced by B^r), will be called a* smooth weak Liapunov function in the large.

Definition 2.7 *A smooth strict Liapunov function in the small is a weak Liapunov function such that $\dot{V}(x)$ is negative definite; in other words, it satisfies, instead of (iv),*
(v) $\nabla V(x) \cdot f(x) < 0$ for each $x \in B_r$ $(x \neq 0)$.

A function $V(x)$ defined for all $x \in \mathbb{R}^n$, which is radially unbounded and fulfills the properties (i), (ii), (iii), (v) with B_r replaced by $\mathbb{R}^n$, will be called a smooth global strict Liapunov function.

---

[3] Although the basic notions are absolutely classical, we warn the reader that the terminology we are going to introduce should be intended for the use of the present book. It aims to emphasize certain features of interest for our purposes and may differ from that more widely used in the literature.

Stability properties can be checked by means of appropriate Liapunov functions, according to the following well known criteria.

Theorem 2.1 *(First Liapunov Theorem) If there exists a smooth weak Liapunov function in the small, then (2.13) is stable at the origin.*

Theorem 2.2 *(Second Liapunov Theorem) If there exists a smooth strict Liapunov function in the small, then (2.13) is locally asymptotically stable at the origin.*

If there exists a smooth global strict Liapunov function, then (2.13) is globally asymptotically stable at the origin.

Theorem 2.3 *If there exists a smooth weak Liapunov function in the large, then (2.13) is Lagrange stable.*

The third theorem is due to Yoshizawa ([159]). The proofs of these theorems are easy and are not reported here.

Remark 2.3 As far as Liapunov functions are assumed to be of class (at least) C^1, condition (iv) is clearly equivalent to the following one:

(iv') for each solution $x(\cdot)$ of (2.13) defined on some interval I and lying in B_r, the composite map $t \mapsto V(x(t))$ is non-increasing on I.

Such a monotonicity condition can be considered as a "nonsmooth analogous" of properties (iii), (iv). Indeed, it can be stated without need of any differentiability (or even continuity) assumption about V.

As we shall see later, assuming that $V(x)$ is continuous and replacing (iv) by (iv'), is actually sufficient in order to prove Theorems 2.1. ■

Remark 2.4 *For certain applications (see for instance [128], [151]), it is not convenient to work with a Liapunov function which is necessarily positive definite. Stability theorems which make use of a semi-definite Liapunov function can be found in [82] and [77]. Of course, in this case one needs to check some additional assumptions.* ■

2.2.2 Converse theorems

From a mathematical point of view, the question whether Theorems 2.1, 2.2 and 2.3 are invertible is quite natural. Recently, it has been recognized to be an important question also for applications to control theory.

Asymptotic stability

Great contributions to studies about the invertibility of second Liapunov Theorem were due to Malkin, Barbashin and Massera, around 1950. In particular, in [98] Massera proved the converse under the assumption that the vector field f is locally Lipschitz. For such vector fields, he proved that asymptotic stability actually implies the existence of a Liapunov function of class C^∞. In 1956, Kurzweil ([91]) proved that the regularity assumption about f can be relaxed. The full statement of Kurzweil's Theorem is postponed to next chapter. For the moment, we can limit ourselves to the following partial version of it.

Theorem 2.4 *Let f be continuous. If (2.13) is locally asymptotically stable at the origin then there exists a C^∞ strict Liapunov function in the small.*

If the system is globally asymptotically stable at the origin, then there exists a C^∞ global strict Liapunov function.

It is worth noticing that Kurzweil's Theorem provides a Liapunov function of class C^∞ in spite of f being only continuous.

Example 2.1 Consider the two-dimensional system

$$\begin{cases} \dot{x}_1 = -x_1^{1/3} \\ \dot{x}_2 = x_1^{2/3} x_2^{1/3} - x_2 \ . \end{cases} \tag{2.14}$$

It is clear that the right hand side is continuous, but not Lipschitz continuous in any neighborhood of the origin. In particular, forward uniqueness fails for initial conditions of the form $(\bar{x}_1, 0)$, and backward uniqueness fails for initial conditions of the form $(0, \bar{x}_2)$. By direct integration, it is possible to see that the origin is globally asymptotically stable. The vector field and some trajectories of system (2.14) are plotted in Figure 2.2. As it is easy to check, an explicit Liapunov function of class C^1 for system (2.14) is given by

$$V(x_1, x_2) = \frac{1}{2} x_1^2 + \frac{3}{4} x_2^{4/3} \ .$$

This example has been taken from the recent paper [110].

Stability

Unfortunately, such a strong result does not hold for Liapunov and Lagrange stability. Indeed, it is well known that there exist stable systems with C^1 right hand side, which have no continuous Liapunov functions.

32   Time invariant systems

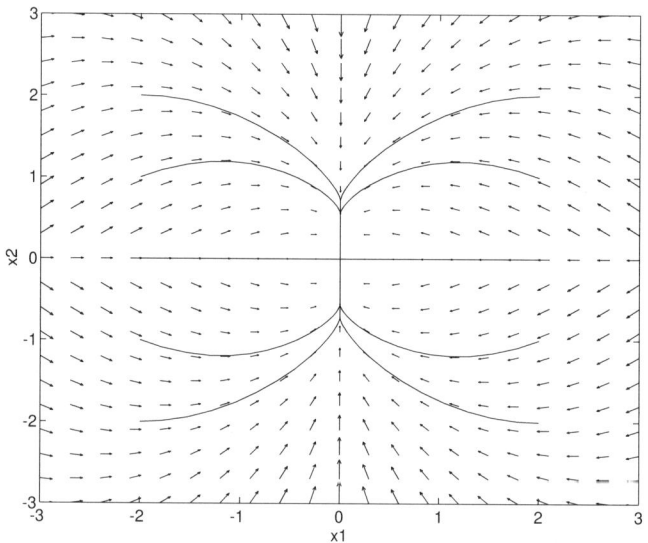

Figure 2.2: System of Example 2.1

Example 2.2 The example is due to Krasovskii ([88], see also [9]). Let us consider the two-dimensional system

$$\begin{cases} \dot{x}_1 = x_2 \\ \dot{x}_2 = -x_1 + x_2 g(x_1^2 + x_2^2) \end{cases}$$

where $g(\xi) = \xi^3 \sin^2(\pi/\xi)$, $g(0) = 0$. It corresponds to the so called *center-focus* configuration. The origin is surrounded by infinitely many limit cycles with common center at the origin and radius which goes to zero. The annular regions between two consecutive limit cycles are covered by spirals. Each solution describing a spiral winds round the internal cycle for $t \to -\infty$, and round the external one for $t \to +\infty$. Note that since f is of class C^1, we have uniqueness of solutions.

It should be clear that this system is stable at the origin. However, continuous Liapunov functions cannot exist. Indeed, let $V(x)$ be a function for which the monotonicity condition (iv') holds. Then, it must be necessarily constant along every limit cycle. Let in addition V be continuous, and let Γ_1, Γ_2 be two consecutive limit cycle (let us agree that Γ_1 lies inside Γ_2). Since Γ_1 and Γ_2 are asymptotically "joined" by trajectories, the value of V on Γ_1 cannot be less than the value of V on Γ_2. By repeating the argument, we conclude that the limit of $V(x)$ as $x \to 0$ coincides with the supremum of V in some neighborhood of the origin. This is clearly impossible because of properties (i) and (ii) of Definition 2.5.

Note that a similar configuration can be obtained by means of vector fields of an arbitrary degree of smoothness, and even of class C^∞. Note also that there is a 1-dimensional version of this example. The simplest way to construct it, is to take

$$\dot{x} = g(x)$$

where g is the same as before. However, in the next chapter we consider a different scalar equation which exhibits an analogous behavior but with the additional advantage that its solutions can be explicitly computed.

We note finally that similar conclusions hold for the case of Lagrange stability. This can be seen by an obvious modification of the present example. ∎

The following examples are new. They show that the lack of continuity is not the only obstruction to the existence of smooth Liapunov functions.

Example 2.3 Let $\psi : \mathbb{R}^2 \to \mathbb{R}$ be a C^∞ function such that $\psi(x_1, x_2) > 0$ on $\{(x_1, x_2) \in \mathbb{R}^2 : x_1 x_2 \neq 0\}$, and $\psi(x_1, x_2)$ vanishes, together with its partial derivatives of any order on $\{(x_1, x_2) \in \mathbb{R}^2 : x_1 x_2 = 0\}$. Consider the vector field of $\mathbb{R}^2$

$$f(x_1, x_2) = \begin{pmatrix} f_1(x_1, x_2) \\ f_2(x_1, x_2) \end{pmatrix} = \psi(x_1, x_2) \begin{pmatrix} \operatorname{sgn} x_1 \\ -\operatorname{sgn} x_2 \end{pmatrix}$$

that is, the system

$$\begin{cases} \dot{x}_1 = f_1(x_1, x_2) \\ \dot{x}_2 = f_2(x_1, x_2) \end{cases} \quad (2.15)$$

whose trajectories are shown in Figure 2.3.

34 Time invariant systems

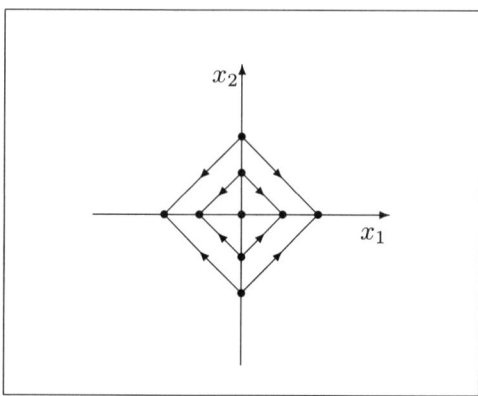

Figure 2.3: Trajectories of the system of Example 2.3

It is clear that $f \in C^\infty$, and that the system is Liapunov stable at the origin and Lagrange stable.

Assume now that there exists a C^1 weak Liapunov function $V(x_1, x_2)$ for (2.15). Then, for each $\epsilon > 0$ there must exist a number $\bar{x}_1$ with $0 < \bar{x}_1 < \epsilon$ such that

$$\frac{\partial V}{\partial x_1}(\bar{x}_1, 0) > 0 \; . \tag{2.16}$$

Otherwise, the function $x_1 \mapsto V(x_1, 0)$ would be non-increasing and since $V(0,0) = 0$ the positive definiteness assumption is contradicted. Let us fix such a point $\bar{x}_1$. For any $x_2 > 0$, we have also

$$\frac{\partial V}{\partial x_1}(\bar{x}_1, x_2)\psi(\bar{x}_1, x_2) - \frac{\partial V}{\partial x_2}(\bar{x}_1, x_2)\psi(\bar{x}_1, x_2) \leq 0 \; .$$

Since $\psi \geq 0$, taking the limit for $x_2 \to 0$ we get

$$\frac{\partial V}{\partial x_1}(\bar{x}_1, 0) - \frac{\partial V}{\partial x_2}(\bar{x}_1, 0) \leq 0 \; . \tag{2.17}$$

On the other hand, for $x_2 < 0$ we have

$$\frac{\partial V}{\partial x_1}(\bar{x}_1, x_2)\psi(\bar{x}_1, x_2) + \frac{\partial V}{\partial x_2}(\bar{x}_1, x_2)\psi(\bar{x}_1, x_2) \leq 0 \; .$$

Arguing as before, this yields

$$\frac{\partial V}{\partial x_1}(\bar{x}_1, 0) + \frac{\partial V}{\partial x_2}(\bar{x}_1, 0) \leq 0 \; . \tag{2.18}$$

Comparing (2.17) and (2.18), we obtain

$$\frac{\partial V}{\partial x_1}(\bar{x}_1, 0) \leq 0$$

which is a contradiction to (2.16).

If we agree to define weak Liapunov functions by requiring (iv') instead of (iv), it is immediate to verify that $V(x_1, x_2) = |x_1| + |x_2|$ is a continuous (actually, globally Lipschitz continuous) weak Liapunov function for (2.15). ∎

Summing up, we discovered that for stable (or Lagrange stable) C^∞ vector fields, we may have no continuous Liapunov functions, or even Lipschitz continuous Liapunov functions but not a smooth one. The next example shows another possibility. There exist stable, C^∞ systems which have continuous Liapunov functions but not a Lipschitz continuous one.

Example 2.4 As in Example 2.3, we consider a planar vector field (2.15) where f is of class C^∞ and vanishes on the coordinate axes. The difference is that now the trajectories lying outside the coordinate axes are arcs of circumferences. More precisely, for any $r > 0$ we have four solutions whose images can be re-parametrized in the following way:

$$\gamma_1 \begin{cases} x_1 = r(\cos\theta + 1) \\ x_2 = r(\sin\theta + 1) \end{cases} \quad \pi < \theta < \frac{3\pi}{2},$$

$$\gamma_2 \begin{cases} x_1 = r(\cos\theta - 1) \\ x_2 = r(\sin\theta + 1) \end{cases} \quad \frac{3\pi}{2} < \theta < 2\pi,$$

$$\gamma_3 \begin{cases} x_1 = r(\cos\theta - 1) \\ x_2 = r(\sin\theta - 1) \end{cases} \quad 0 < \theta < \frac{\pi}{2},$$

$$\gamma_4 \begin{cases} x_1 = r(\cos\theta + 1) \\ x_2 = r(\sin\theta - 1) \end{cases} \quad \frac{\pi}{2} < \theta < \pi.$$

We emphasize that the re-parametrizations preserve the orientation of the trajectories.

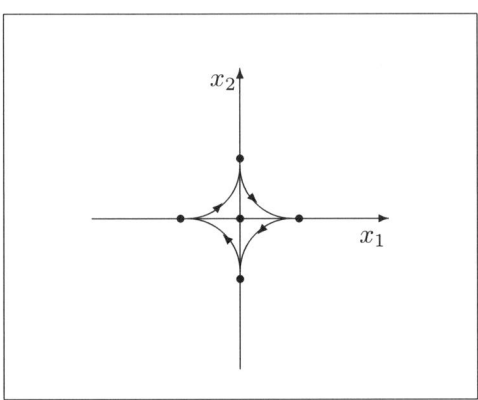

Figure 2.4: Trajectories of the system of Example 2.4

The system is stable at the origin (and Lagrange stable). A continuous Liapunov function can be obtained by noticing that for fixed $r > 0$, $\gamma = \overline{\gamma_1} + \overline{\gamma_2} + \overline{\gamma_3} + \overline{\gamma_4}$ is a closed curve which surrounds the origin. So, we can set $V(x) = r$ for $x \in \text{Im}\gamma$.

Assume now that there exists a locally Lipschitz continuous Liapunov function $W(x_1, x_2)$. It follows that also the function $r \mapsto W(r, 0)$ is locally Lipschitz continuous, and hence a.e. differentiable. Pick any $r > 0$ such that $\frac{\partial W}{\partial x_1}(r, 0)$ exists, and look at the curve γ_1, for which we have

$$\gamma_1'(r) = \begin{pmatrix} -r\sin\theta \\ r\cos\theta \end{pmatrix}.$$

Fix further $R > r$, and let L be a Lipschitz constant for W on the region $\{(x_1, x_2) : |x_1| + |x_2| \leq R\}$.

Since W is non-increasing along the trajectories, we easily get that $W(\gamma_1(\theta))$ is non-increasing, as well. It follows that

$$\limsup_{\theta \to \frac{3\pi}{2}^-} \frac{W(\gamma_1(\theta)) - W(\gamma_1(\frac{3\pi}{2}))}{\theta - \frac{3\pi}{2}} \leq 0.$$

On the other hand

$$\frac{W(\gamma_1(\theta)) - W(\gamma_1(\frac{3\pi}{2}))}{\theta - \frac{3\pi}{2}} = \frac{W(\gamma_1(\theta)) - W(\gamma_1(\frac{3\pi}{2}) + (\theta - \frac{3\pi}{2}) \cdot \gamma_1'(\frac{3\pi}{2}))}{\theta - \frac{3\pi}{2}}$$

$$+ \frac{W(\gamma_1(\frac{3\pi}{2}) + (\theta - \frac{3\pi}{2}) \cdot \binom{r}{0}) - W(\gamma_1(\frac{3\pi}{2}))}{\theta - \frac{3\pi}{2}}.$$

But

$$\left|\frac{W(\gamma_1(\theta)) - W(\gamma_1(\frac{3\pi}{2}) + (\theta - \frac{3\pi}{2}) \cdot \gamma_1'(\frac{3\pi}{2}))}{\theta - \frac{3\pi}{2}}\right|$$

$$\leq L \left|\frac{\gamma_1(\theta) - \gamma_1(\frac{3\pi}{2})}{\theta - \frac{3\pi}{2}} - \gamma_1'(\frac{3\pi}{2})\right| = o(1)$$

as $\theta \to \frac{3\pi}{2}^-$ and

$$\frac{W(\gamma_1(\frac{3\pi}{2}) + (\theta - \frac{3\pi}{2}) \cdot \binom{r}{0}) - W(\gamma_1(\frac{3\pi}{2}))}{\theta - \frac{3\pi}{2}}$$

$$= \frac{W(r + r(\theta - \frac{3\pi}{2}), 0) - W(r, 0)}{\theta - \frac{3\pi}{2}} \to r \frac{\partial W}{\partial x_1}(r, 0)$$

as $\theta \to \frac{3\pi}{2}^-$. We infer that $\frac{\partial W}{\partial x_1}(r,0) \leq 0$ for a.e. $r > 0$, hence the locally Lipschitz continuous map $r \mapsto W(r,0)$ is non-increasing and $W(r,0) \leq 0$ for $r > 0$, which is impossible because of the definition of Liapunov function. ∎

Converse theorems which guarantee the existence of smooth Liapunov functions for Liapunov and Lagrange stability are known only in the one-dimensional case (see [17]) and, of course, in the linear case.

2.2.3 Generalized Liapunov functions

The examples of the previous subsection motivate the need for a more general definition of Liapunov function, at least for the weak versions.

Definition 2.8 *Let $r > 0$. A function $V : B_r \to \mathbb{R}$ is called a generalized weak Liapunov function in the small if it satisfies (i), (iv') and, in addition, the following two properties:*

(ii') for some $\eta < r$ and for each $\sigma \in (0, \eta)$ there exists $\lambda > 0$ such that $V(x) > \lambda$ when $\sigma \leq ||x|| \leq \eta$

(iii') $V(x)$ is continuous at $x = 0$.

Of course, (ii') implies that $V(x)$ is positive definite. On the other hand, it is easy to see that (ii') is satisfied if $V(x)$ is, for instance, lower semi-continuous and positive definite. The definition of weak generalized Liapunov function in the large is similar. The following theorem is due to Auslander and Seibert ([9]).

38   Time invariant systems

Theorem 2.5 *System (2.13) is Liapunov stable at the origin if and only if there exists a generalized weak Liapunov function in the small.*

Proof. Let us prove first the sufficient part. Fix $\varepsilon > 0$ and assume, without loss of generality, that $\varepsilon < \eta$. According to (ii'), we can find $\lambda > 0$ such that $V(x) > \lambda$ for $\varepsilon \leq \|x\| \leq \eta$. Since $V(0) = 0$ and $V(x)$ is continuous at $x = 0$, there exists $\delta > 0$ such that $V(x) < \lambda$ for $\|x\| \leq \delta$.

Now, pick any x_0 with $\|x_0\| < \delta$, and assume that there exist a solution $x(\cdot) \in \mathcal{S}_{x_0}$ and a time $t_2 > 0$ such that $\|x(t_2)\| \geq \varepsilon$. Since $x(\cdot)$ is continuous and $\delta < \varepsilon$, we must have $\|x(t_1)\| = \varepsilon$ for some $t_1 \in (0, t_2)$. Without loss of generality, we can also assume that t_1 is the smallest time for which this happens, so that $\|x(t)\| < \varepsilon < r$ for all $t \in [0, t_1)$.

We conclude that $V(x(0)) < \lambda$, while $V(x(t_1)) > \lambda$ with $t_1 > 0$. A contradiction to (iv'). Thus we see that $x(\cdot)$ cannot leave the ball of radius ε, as required by the definition of stability. In particular, $x(\cdot)$ is right continuable to the infinity.

As far as the necessary part is concerned, let us define

$$V(\xi) = \sup\{\|x(t)\| : t \geq 0, x(\cdot) \in \mathcal{S}_\xi\} .$$

We prove that such a function satisfies all the required properties. According to the stability assumption, there exists $r > 0$ such that $V(x) < +\infty$ for $x \in B_r$. It is obvious that $V(x) \geq \|x\|$ if $0 \neq x \in B_r$, and this in turn implies (ii'). It is also clear that $V(0) = 0$ because of stability. The stability assumption is invoked also in order to prove that $V(x)$ is continuous at the origin. Indeed, $\forall \varepsilon > 0, \exists \delta > 0$ such that

$$\|\xi\| < \delta, \ x(\cdot) \in \mathcal{S}_\xi \implies \|x(t)\| < \varepsilon, \ t \geq 0$$

and hence $V(\xi) \leq \varepsilon$.

The monotonicity condition (iv') is a trivial consequence of the definition and the fact that new solutions can be obtained piecing together solutions defined on consecutive intervals. ∎

We report also the following results, due to Yorke ([157]).

Theorem 2.6 *Assume that the right hand side of (2.13) is locally Lipschitz continuous. Then, if (2.13) is Liapunov stable at the origin there exists a lower semi-continuous generalized weak Liapunov function in the small.*

Proof. Define $V(x)$ as in the proof of the previous theorem. It remains to prove that V is lower semicontinuous at an arbitrary point $x_0 \in B_r$. Being f locally Lipschitz continuous, uniqueness of solutions and continuous dependence with respect to the initial values are guaranteed.

By the definition of V, for each $\varepsilon > 0$ there is $\tau > 0$ such that

$$V(x_0) - \varepsilon/2 < \|x(\tau; x_0)\| \ .$$

Moreover, there is $\delta > 0$ such that for all ξ with $\|\xi - x_0\| < \delta$ one has

$$\|x(\tau; x_0)\| - \|x(\tau; \xi)\| \le \|x(\tau; x_0) - x(\tau; \xi)\| < \varepsilon/2 \ .$$

Hence,

$$V(x_0) - \varepsilon/2 < \|x(\tau; x_0)\| \le \|x(\tau; \xi)\| + \varepsilon/2 \ .$$

Again, by the definition of V, we have $\|x(\tau; \xi)\| \le V(\xi)$. In conclusion,

$$V(x_0) - \varepsilon < \|x(\tau; \xi)\| \le V(\xi)$$

that was required to prove. ∎

Remark 2.5 Under the same assumptions, it is also possible to construct an upper semi-continuous Liapunov function, setting

$$\tilde{V}(\zeta) = \inf\{\|x(t; \zeta)\| : t \le 0\} \ .$$

Of course, $\tilde{V}(x) \le V(x)$ for each x.

Remark 2.6 Theorems 2.5 and 2.6 refer to local stability, but analogous results can be formulated for Lagrange stability.

2.2.4 Absolute stability

Auslander and Seibert ([9]) discovered also that the existence of a generalized Liapunov function continuous in a whole neighborhood of the origin, is equivalent to a stronger form of stability, the so called *absolute stability*. In order to illustrate the idea, we find it convenient to begin with some intuitive considerations. Roughly speaking, stability is a way to describe the behavior of a system in presence of small perturbations of the initial state. Assume now that perturbations are also allowed at arbitrary positive times: under the effect of such perturbations, the system may jump from the present trajectory to a nearby one. Now, it may happen that an unfortunate superposition of these jumps results in an unstable behavior even if the system is stable and the amplitude of the perturbations tends to zero. In fact, this is actually what occurs in the center-focus configuration (Example 2.2).

This phenomenon is technically described by the notion of *prolongation*, due to T. Ura and deeply studied in [9]. As we shall see, the existence of a continuous Liapunov function actually prevents the unstable behavior of prolongational sets. On the other hand, the possibility of taking under control the growth of the prolongational sets leads to the desired strengthened notion of stability.

Throughout this section, we limit ourselves to systems of the form (2.13), whose right hand side is locally Lipschitz continuous, so that uniqueness of solutions is guaranteed. Let us recall that in the topological setting, useful informations about the dynamical behavior of a system can be deduced from the inspection of certain sets. These sets depend in general on the initial state. Thus, they can be reviewed as set valued maps. The simplest example is the *positive orbit* of a point x_0 relative to (2.13), namely

$$\Gamma^+(x_0) = \{y \in \mathbb{R}^n : y = x(t; x_0) \text{ for some } t \geq 0\} .$$

Another example is the positive limit set. We adopt the following agreements about notation. Let $Q(x)$ be a set valued map from $\mathbb{R}^n$ to $\mathbb{R}^n$. For $\Omega \subset \mathbb{R}^n$, we denote $Q(\Omega) = \cup_{x \in \Omega} Q(x)$. Powers of Q will be defined iteratively:

$$Q^0(x) = Q(x) \quad \text{and} \quad Q^k(x) = Q(Q^{k-1}(x))$$

for $k = 1, 2, \ldots$. Moreover, we need to introduce two operators, denoted by $\mathcal{D}$ and $\mathcal{I}$, acting on set valued maps. They are defined according to

$$(\mathcal{D}Q)(x) = \cap_{\delta > 0} \overline{Q(B_\delta(x))}$$

$$(\mathcal{I}Q)(x) = \cup_{k=0,1,2,\ldots} Q^k(x) .$$

The following characterizations are straightforward.

Proposition 2.8 Let $Q(x)$ be any set valued map from $\mathbb{R}^n$ to $\mathbb{R}^n$.

a) $y \in (\mathcal{D}Q)(x)$ if and only if there exist sequences $x_k \to x$ and $y_k \to y$ such that $y_k \in Q(x_k)$ for each $k = 1, 2, \ldots$.

b) $y \in (\mathcal{I}Q)(x)$ if and only if there exists a finite sequence of points $x_0, \ldots, x_K$ such that $x_0 = x$, $y = x_K$ and $x_k \in Q(x_{k-1})$ for $k = 1, 2, \ldots, K$.

The operators $\mathcal{D}$ and $\mathcal{I}$ are idempotent. Moreover, for every set valued map Q and any x the set $(\mathcal{D}Q)(x)$ is closed. However, $(\mathcal{I}Q)(x)$ is not closed in general, not even if $Q(x)$ is closed for each x. When $\mathcal{I}Q = Q$ we say that Q is *transitive*. The positive trajectory is an example of a transitive map. In general, $\mathcal{D}Q$ is not transitive, not even if Q is transitive. Hence, the construction

$$(\mathcal{D}(\ldots(\mathcal{I}(\mathcal{D}(\mathcal{I}(\mathcal{D}Q))))\ldots))(x) \tag{2.19}$$

gives rise to larger and larger sets.

Example 2.5 Consider a one-dimensional equation (2.13), whose dynamics is described in the following way. All the points of the form $P_i = i \in \mathbb{N}$ are equilibria. Moreover, for each $i \geq 1$, there is a monotonically increasing sequence of equilibria $\{P_{i,j}\}_{j=1,2,\ldots}$ such that $P_{i-1} < P_{i,j} < P_i$ for each i and j, and $\lim_j P_{i,j} = P_i$. When x does not belong to the set of points of the type P_i or $P_{i,j}$, we set $f(x) > 0$. In other words, if x is not an equilibrium then under the action of the system it moves to the right.

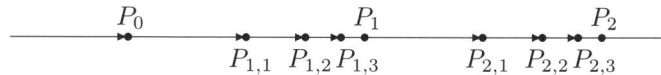

Figure 2.5: Equilibrium points for the system of Example 2.5

It is easy to recognize the following facts:

$$P_{i,1} \in (\mathcal{D}\Gamma^+)(P_{i-1}) \text{ and } P_{i,j} \in (\mathcal{D}\Gamma^+)(P_{i,j-1})$$
$$[P_{i-1}, P_i) = (\mathcal{I}(\mathcal{D}\Gamma^+))(P_{i-1})$$
$$P_i \in (\mathcal{D}(\mathcal{I}(\mathcal{D}\Gamma^+)))(P_{i-1})$$

for $i = 1, 2, \ldots$ and $j = 1, 2, \ldots$. ∎

We are now ready to give the definition.

Definition 2.9 A prolongation *associated to system (2.13) is a set valued map $Q(x)$ which fulfills the following properties:*

(i) for each $x \in \mathbb{R}^n$, $\Gamma^+(x) \subseteq Q(x)$

(ii) $(\mathcal{D}Q)(x) = Q(x)$

(iii) If A is a compact subset of $\mathbb{R}^n$ and $x \in A$, then either $Q(x) \subseteq A$, or $Q(x) \cap \partial A \neq \emptyset$.

If Q is a prolongation and it is transitive, it is called a *transitive prolongation*. The following proposition will be used later (see [9]).

Proposition 2.9 *Let K be a compact subset of $\mathbb{R}^n$ and let Q be a transitive prolongation. Then $Q(K) = K$ if and only if K possesses a fundamental system of compact neighborhoods $\{K_i\}$ such that $Q(K_i) = K_i$.*

Starting from the map Γ^+ and using repeatedly the operators $\mathcal{D}$ and $\mathcal{I}$, we can construct several prolongational sets associated to (2.13). For instance, it is not difficult to see that

$$D_1(x) \underset{\text{def}}{=} (\mathcal{D}\Gamma^+)(x)$$

is a prolongation, the so called *first prolongation* of (2.13). The first prolongation characterizes stability. Indeed, it is possible to prove that the origin is stable for (2.13) if and only if $D_1(0) = \{0\}$.

The intuitive construction (2.19) can be formalized by means of transfinite induction. This allows us to speak about higher order prolongations. More precisely, let α be an ordinal number and assume that the prolongation $D_\beta(x)$ of order β has been defined for each ordinal number $\beta < \alpha$. Then, we set

$$D_\alpha(x) = (\mathcal{D}(\cup_{\beta<\alpha}(\mathcal{I}D_\beta)))(x) .$$

The procedure saturates when $\alpha = \gamma$, the first uncountable ordinal number. Indeed, it is possible to prove that $\mathcal{I}D_\gamma = D_\gamma$, which obviously implies $D_\alpha(x) = D_\gamma(x)$ for each $\alpha \geq \gamma$.

Since, as already mentioned, (2.13) is stable at the origin if and only if $D_1(0) = \{0\}$, it is natural to give the following definition.

Definition 2.10 *Let α be an ordinal number. The origin is stable of order α (or α-stable) for (2.13) if $D_\alpha(0) = \{0\}$. The origin is said to be absolutely stable when it is γ-stable.*

The main result in [9] is as follows.

Theorem 2.7 *The origin is absolutely stable for (2.27) if and only if there exists a generalized weak Liapunov function which is continuous in a whole neighborhood of the origin.*

Proof. First we give a sketch of the proof of the sufficient part. Let $V : B_r \to \mathbb{R}$ be a weak Liapunov function: assume that it is continuous on B_r and that $\liminf_{||x|| \to r^-} V(x) > 0$. For each positive λ, we denote by W_λ the level set $\{x : V(x) \leq \lambda\}$. Moreover, let $\lambda_0 \in (0, \liminf_{||x|| \to r^-} V(x))$. Then, the family of sets $\{W_\lambda\}_{0 < \lambda < \lambda_0}$ is a fundamental system of invariant, compact neighborhoods of the origin. According to Proposition 2.9, it is sufficient to prove that $D_\gamma(W_\lambda) = W_\lambda$ for each $\lambda < \lambda_0$.

Let $x \in W_\lambda$ and $y \in D_1(x)$. By Proposition 2.8, there exist sequences $\{x_k\}$, $\{t_k\}$ such that

$$\lim_{k \to \infty} x(t_k; x_k) = y$$

where $x_k \to x$ and $t_k \geq 0$. For sufficiently large k, we have $x_k \in W_{\lambda_0}$ so that, according to (iv'),

$$V(x(t_k; x_k)) \leq V(x_k) .$$

Using the continuity, we conclude that $V(y) \leq V(x)$, from which it follows $D_1(W_\lambda) = W_\lambda$. The argument is easily completed by transfinite induction.

In order to prove the necessary part, let us assume that the origin is absolutely stable. The idea is to construct the Liapunov function by assigning its level sets for all numbers of the form $j/2^k$ ($k = 0, 1, \ldots, j = 1, \ldots, 2^k$), the so called dyadic rationals. Note that numbers of this type are dense in $[0, 1]$.

Since D_γ is a transitive prolongation, according to Proposition 2.9 we can find a sequence of compact neighborhoods of the origin, labeled as $W_{\frac{1}{2^k}}$ ($k = 0, 1, \ldots$), such that

$$D_\gamma(W_{\frac{1}{2^k}}) = W_{\frac{1}{2^k}}, \quad W_{\frac{1}{2^{k+1}}} \subset \text{Int } W_{\frac{1}{2^k}}, \quad \text{and} \quad \bigcap_k W_{\frac{1}{2^k}} = \{0\}$$

for each $k = 0, 1, \ldots$. Using again Proposition 2.9, we can now find a compact neighborhood of $W_{\frac{1}{2}}$, labelled $W_{\frac{3}{4}}$, such that $D_\gamma(W_{\frac{3}{4}}) = W_{\frac{3}{4}}$, and $W_{\frac{3}{4}} \subset \text{Int } W_1$. By iterating the procedure, we then find $W_{\frac{3}{8}}$, $W_{\frac{5}{8}}$ and $W_{\frac{7}{8}}$ in such a way that

$$W_{\frac{1}{4}} \subset W_{\frac{3}{8}} \subset W_{\frac{1}{2}} \subset W_{\frac{5}{8}} \subset W_{\frac{3}{4}} \subset W_{\frac{7}{8}} \subset W_1$$

and so on. In this way, we arrive to construct W_λ for each dyadic ration λ. Note that if $\lambda' < \lambda''$, then $W_{\lambda'} \subset \text{Int } W_{\lambda''}$.

We are now ready to define $V(x)$ as

$$V(x) = \inf\{\lambda : x \in W_\lambda\}$$

for all $x \in W_1$. This function satisfies all the required properties. In particular, it is clear that $V(0) = 0$ and $V(x) > 0$ for $x \neq 0$. To see that $V(x)$ is not increasing along the solutions, let us fix $x_0 \in W_1$ and assume that $V(x(t;x_0)) > V(x_0)$ for some $t \geq 0$. Let λ be a dyadic rational such that $V(x(t;x_0)) > \lambda > V(x_0)$, so that $x_0 \in W_\lambda$. Since $D_\gamma(W_\lambda) = W_\lambda$, we must have in particular $x(t;x_0) \in W_\lambda$. This implies $V(x(t;x_0)) \leq \lambda$. A contradiction.

We finally show that V is continuous on $\text{Int } W_1$. Assume the contrary. Then for some $\bar{x} \in \text{Int } W_1$ there exists a sequence $x_\nu \to \bar{x}$ such that $V(x_\nu)$ does not converge to $V(\bar{x})$. Since V is bounded, by possibly taking a subsequence, we can assume that $\lim_\nu V(x_\nu) = l \neq V(\bar{x})$.

Consider first the case $V(\bar{x}) < l$, and pick a dyadic rational λ in such a way that $V(\bar{x}) < \lambda < l$ and $\bar{x} \in \text{Int } W_\lambda$. For every sufficiently large ν, we should have $x_\nu \in W_\lambda$ as well. But then, $V(x_\nu) \leq \lambda$, and this is a contradiction.

The case $V(\bar{x}) > l$ is treated in a similar way. ∎

Note that the constructions of the Liapunov function in Theorems 2.5 and 2.7 are quite different.

2.2.5 Stability and robustness

As already noticed, the meaning of Liapunov stability can be described as a sort of robustness with respect to perturbations of the initial state. In general, a stable system is not robust with respect to perturbations acting on the vector field f. Rather, robustness with respect to permanently acting disturbances is a typical property of systems which exhibit stronger stability features, such as asymptotic stability. In order to recognize it, the method of Liapunov functions turns out to be the more direct approach.

There are many possible ways to formalize this kind of robustness. The following definition goes back to J.G. Malkin (see [66], Sect. 6, for other references to the Russian literature). Here, together with (2.13), we need to consider another system

$$\dot{x} = g(x), \qquad x \in \mathbb{R}^n . \tag{2.20}$$

We adopt the symbol $\mathcal{S}^g_{x_0}$ to denote the set of solutions of (2.20) issuing from x_0. As usual, we locate the equilibrium position at the origin.

Definition 2.11 *System (2.13) is said to be* totally stable *at the origin if for each $\varepsilon > 0$ there exist two numbers $\delta_1 > 0$ and $\delta_2 > 0$ such that for each system of the form (2.20) and each initial state x_0 one has*

$$\left.\begin{array}{l} \|f(x) - g(x)\| < \delta_1, \quad \text{for } \|x\| < \varepsilon \\ \|x_0\| < \delta_2 \end{array}\right\} \implies \|x(t)\| < \varepsilon$$

for all $x(\cdot) \in \mathcal{S}^g_{x_0}$ and each $t \geq 0$.

Malkin proved that if (2.13) is asymptotically stable at the origin, then it is totally stable. The converse is false, at least for $f \in C^\infty$ (a counter-example can be obtained by an easy modification of previous Example 2.2, see again [66]): at the best knowledge of the authors, it is not known if the converse holds in the analytic case.

Note that the concept of total stability does not say that if (2.13) is stable then close to f there are vector fields g for which (2.20) possesses some kind of stability property. For instance, the scalar equation $\dot{x} = -x^3$ is asymptotically stable, and hence totally stable, but the equation $\dot{x} = k^2 x - x^3$ is unstable, no matter how small k is.

A different type of notion can be obtained if the number δ_1 in Definition 2.11 is allowed to vary with the point x (and, in particular, to become smaller and smaller as x tends to zero).

Theorem 2.8 *Let the origin be an asymptotically stable equilibrium position for system (2.13). Then, there exist $\varepsilon_0 > 0$ and a positive definite continuous function $\delta(x) : B_{\varepsilon_0} \to \mathbb{R}$ such that the following holds: if $g(x)$ is any vector field satisfying the conditions:*

- *$g(x)$ is continuous*
- *$g(0) = 0$*
- *$\|f(x) - g(x)\| < \delta(x)$ for each $x \in B_{\varepsilon_0}$*

then the origin is also an asymptotically stable equilibrium position for (2.20).

Proof. According to Kurzweil's converse theorem, we can find a strict Liapunov function $V(x)$ of class at least C^1, defined in a closed neighborhood of the origin $\overline{B_{\varepsilon_0}}$. Let $\nabla V(x) \cdot f(x) = -W(x)$, so that $W(x) > 0$ for $x \in B_{\varepsilon_0}$ ($x \neq 0$). Let $\alpha(x)$ be any strictly positive, continuous function (possibly constant) such that

$$\|\nabla V(x)\| \le \alpha(x)$$

for $x \in B_{\varepsilon_0}$. Let finally $\delta(x) = W(x)/(2n\alpha(x))$.

Now, assume that $g(x)$ satisfies the required conditions. If we compute the derivative of $V(x)$ with respect to (2.20) we obtain

$$\begin{aligned} \nabla V(x) \cdot g(x) &= \nabla V(x) \cdot [g(x) - f(x)] + \nabla V(x) \cdot f(x) \\ &\le n\|\nabla V(x)\| \cdot \|g(x) - f(x)\| - W(x) \le -\frac{W(x)}{2} \end{aligned}$$

for each $x \in B_{\varepsilon_0}$ (recall that we are using the sup-norm). The conclusion is implied by second Liapunov theorem. ∎

Disturbances permanently acting on the vector field are also called *outer perturbations* [4].

In view of practical applications, it is very important to take into consideration also the effect of perturbations due to state measurement errors.

A differential equation like (2.13) can be thought of as rule which supplies the value of the velocity $\dot{x}$ when the state x is known. But assume that the state cannot be measured in exact way: at some instant t, the system is actually at the state x, but we believe that the state is $x + s(x)$, where $s(x)$ is some (hopefully small) continuous function. Accordingly, we are led to assume for the velocity $\dot{x}$ at the instant t the value $f(x + s(x))$ instead of the correct one $f(x)$. The computed solution will be therefore affected by errors. In this situation, we say that the system is subject to *inner perturbations*. If f is continuous, inner perturbations can be easily reduced to outer perturbations, by simply writing

$$f(x + s(x)) = f(x) + [f(x + s(x)) - f(x)] .$$

Theorem 2.9 *Let the origin be an asymptotically stable equilibrium position for system (2.13). Then, there exist $\varepsilon_0 > 0$ and a positive definite continuous function $\sigma(x) : B_{\varepsilon_0} \to \mathbb{R}$ such that the following holds: if $s(x)$ is any vector field satisfying the conditions:*

- *$s(x)$ is continuous*

---

[4] We point out that for more generality, outer perturbations are often represented in the literature by means of functions of time.

- $s(0) = 0$
- $\|s(x)\| < \sigma(x)$ for each $x \in B_{\varepsilon_0}$

then the origin is also an asymptotically stable equilibrium position for the system

$$\dot x = f(x + s(x)) \,.$$

Global versions of Theorems 2.8 and 2.9 can be obtained by some obvious modifications of the previous statements.

We remark that the treatment of stability under inner and outer perturbations is much more delicate in the case of discontinuous right hand side. We do not go into this subject, but we point out some classical work ([69]) and some recent interesting research papers ([94], and the references therein).

2.2.6 The Invariance Principle

Converse Liapunov Theorem 2.4 about asymptotic stability guarantees the existence of some strict Liapunov functions, but does not give any general indication on how to construct one of them. In fact, experience shows that in general finding explicit Liapunov functions is far from being trivial, even when one is sure that they exist; from a practical point of view, this is the main drawback of Liapunov method.

In this subsection we discuss a way to ascertain asymptotic stability which can be of some advantage when only a weak Liapunov function is available. The main tool is provided by the well known *Invariance Principle*, attributed to Barbashin and Krasowski in the Russian literature, and to LaSalle in the western literature ([67], [93]).

Let us consider again the time invariant system of ordinary differential equations (2.13), under the assumption that the vector field f is locally Lipschitz continuous, so that for each initial state x_0 there is just a unique solution $x(t; x_0)$. Let x_0 be a point such that $x(t; x_0)$ is defined for all $t \geq 0$. A point y is said to be a *limit point* of x_0 if there exists a sequence $\{t_k\}$ such that $\lim_k t_k = +\infty$ and $\lim_k x(t_k; x_0) = y$. The set of the limit points of x_0 is denoted $\Lambda^+(x_0)$ and it is called the *(positive) limit set* of x_0.

Lemma 2.2 ([20]) *Assume that the solution $x(t; x_0)$ is bounded for $t \geq 0$. Then, $\Lambda^+(x_0)$ is nonempty, compact and invariant (that is, if $y \in \Lambda^+(x_0)$ then $x(t; y) \in \Lambda^+(x_0)$ for each $t \in \mathbb{R}$).*

Theorem 2.10 *(Invariance Principle)* Let Ω be an open set containing the origin of $\mathbb{R}^n$. Let $V(x)$ be a smooth weak Liapunov function for the system (2.13), defined for each $x \in \Omega$ and satisfying the conditions (i), (ii), (iii), (iv) of Definition 2.5 for each $x \in \Omega$. Let $l > 0$ and let Ω_l be the connected component of the level set $\{x \in \Omega : V(x) \leq l\}$ which contains the origin. Assume that Ω_l is compact, and let

$$E = \{x \in \Omega_l : \dot{V}(x) = 0\} \ .$$

Let finally K_0 be the union of all the sets M which are compact and invariant with respect to (2.13) and which are contained in E.

Then, for each $x_0 \in \Omega_l$, we have $\Lambda^+(x_0) \subseteq K_0$.

Proof Since there exists a weak Liapunov function, the origin is stable. In particular, it is an equilibrium position and constitutes a compact and invariant set. Clearly $\dot{V}(0) = 0$. It follows that $0 \in K_0$, so that K_0 is nonempty.

First of all, we remark that if $x_0 \in \Omega_l$, then $x(t; x_0) \in \Omega_l$ for each $t \geq 0$. This can be proven by contradiction. Let us assume that for some $t_1 > 0$, $x(t_1; x_0) \notin \Omega_l$. If $V(x(t_1; x_0)) > l$, then the monotonicity condition is violated. If instead $V(x(t_1; x_0)) \leq l$, then $x(t_1; x_0)$ belongs to some other connected component of the level set. Hence, there exists $t_2 \in (0, t_1)$ for which $V(x(t_2; x_0)) > l$. The condition follows again by the monotonicity condition.

Since Ω_l is compact, $x(t; x_0)$ is bounded for $t \geq 0$. By Lemma 2.2, $\Lambda^+(x_0)$ is nonempty and compact. Moreover, $\Lambda^+(x_0) \subseteq \Omega_l$. Using again the monotonicity condition, we see that the limit $l_0 = \lim_{t \to +\infty} V(x(t; x_0))$ exists. In addition we have $0 \leq l_0 \leq l$. The definition of limit set implies that $V(y) = l_0$ for any $y \in \Lambda^+(x_0)$.

Now, using the invariance property, we also see that $V(x(t; y)) = l_0$ for each $t \in \mathbb{R}$. Finally, a simple computation shows that $\dot{V}(y) = 0$.

We have so proved that $\Lambda^+(x_0) \subseteq E$. The conclusion easily follows. ∎

Roughly speaking, the Invariance Principle states that the trajectory issuing from each $x_0 \in \Omega_l$ is attracted by K_0. Of course, under the assumptions of Theorem 2.10 the origin is stable. Thus, if one is able to exclude the existence of nontrivial invariant subsets of E, the asymptotic stability of the origin is proven. Sometimes, this can be done on the basis of simple geometric considerations.

Example 2.6 The linear system

$$\begin{cases} \dot{x}_1 = -x_1 - x_2 \\ \dot{x}_2 = x_1 \ . \end{cases}$$

is asymptotically stable at the origin. It admits $V(x_1, x_2) = x_1^2 + x_2^2$ as a weak Liapunov function. The set $\Omega_l = \{(x_1, x_2) : x_1^2 + x_2^2 \leq l\}$ is compact for each $l > 0$, and $E = \{(x_1, x_2) : x_1 = 0\} \cap \Omega_l$. At a point of the form $(0, x_2)$, with $x_2 \neq 0$, the vector field is orthogonal to the x_2-axis. Thus, it is clear that no trajectory of the system (apart from the trivial one) is contained in E. Hence, K_0 reduces to the origin, and Theorem 2.10 allows us to get the conclusion.

Of course, in this case it is not difficult to find also a strict Liapunov function (for instance, $V(x_1, x_2) = 3x_1^2 + x_1 x_2 + 2x_2^2$). ∎

A version of the Invariance Principle for systems without uniqueness of solutions can be found in [60].

2.2.7 The domain of attraction

As in the previous subsection, we consider the time invariant system of ordinary differential equations (2.13), and we assume that the vector field f is locally Lipschitz continuous, so that uniqueness of the solution for each initial state is guaranteed. Moreover, let us assume that the origin is an asymptotically stable equilibrium point. The set

$$\mathcal{A} = \{x_0 \in \mathbb{R}^n : \lim_{t \to +\infty} x(t; x_0) = 0\}$$

is called the *domain of attraction*. Clearly, the origin is globally asymptotically stable if, and only if, $\mathcal{A} = \mathbb{R}^n$. If it is not the case, it is important for practical reasons to have information about the size and/or the shape of $\mathcal{A}$. Indeed, the stability properties could be of scarce utility if the domain of attraction is very small, or if the origin is very close to its boundary. There is a wide literature about theoretical methods for the determination of $\mathcal{A}$, and about numerical methods for its approximate estimation (we refer to [65] for a very good survey, covering the attempts made until 1985; we refer also to the recent paper [28]). We start by recalling some geometric properties of $\mathcal{A}$.

Proposition 2.10 *Let the origin be (locally) asymptotically stable for (2.13). Then $\mathcal{A}$ and $\partial \mathcal{A}$ are invariant with respect to the trajectories of (2.13). Moreover, $\mathcal{A}$ is diffeomorphic to $\mathbb{R}^n$. In particular, $\mathcal{A}$ is open and simply connected.*

The most celebrated approach to the characterization of the domain of attraction is based on the following statement ([163]; see also [67] and [153]).

Theorem 2.11 *Let A be an open subset of $\mathbb{R}^n$, such that $0 \in A$. Let us assume that there exist two functions $W : A \to [0,1)$ and $h : A \to [0, +\infty)$ such that:*

(i) h is continuous and W is of class C^1;

(ii) $h(0) = 0$, and $h(x) > 0$ for each $x \in A$ ($x \neq 0$);

(iii) $W(0) = 0$, and $W(x) > 0$ for each $x \in A$ ($x \neq 0$);

(iv) $\nabla W(x) \cdot f(x) = -h(x)(1 - W(x))$ for each $x \in A$;

(v) $\lim_{k \to \infty} W(x_k) = 1$ for each sequence $\{x_k\}$ such that $x_k \in A$ for $k = 1, 2, \ldots$, and either $x_k \to \bar{x} \in \partial A$ or $||x_k|| \to \infty$.

Then, A coincides with the domain of attraction of (2.13).

Condition (iv) of the previous theorem can be interpreted as a partial differential equation of the Hamilton-Jacobi type in the unknown $W(x)$, with an arbitrary term inside (the positive definite function h). This equation is usually called *Zubov's equation*. Finding explicit solutions of Zubov's equation is very hard in general, though an appropriate choice of h may be of some help.

To clarify the link with the Liapunov method, we point out that the substitution $W = 1 - e^{-V}$ put Zubov's equation in the equivalent form

$$\nabla V(x) \cdot f(x) = -h(x) \ .$$

Accordingly, the boundary condition (v) becomes $\lim_{k \to \infty} V(x_k) = \infty$.

Zubov's equation appears in [163] and [67] in the more involved form

$$\nabla W(x) \cdot f(x) = -h(x)(1 - W(x))\sqrt{1 + |f(x)|^2} \ .$$

The new multiplicative term is due to a time reparametrization $ds = \sqrt{1 + |f(x)|^2}\, dt$, which may be motivated in general by the need of preventing finite escape time of solutions. We can limit ourselves to the simpler form (iv) since the continuability of solutions is implicitly guaranteed a priori by the regularity of $f(x)$ and the stability assumption.

Theorem 2.11 admits a converse (see [153]), which can be stated in the following form.

Theorem 2.12 *Let the vector field $f(x)$ be of class C^1, and assume that the origin is an asymptotically stable equilibrium point with domain of attraction A. Then, there exist functions $V : A \to [0, +\infty)$ and $h : A \to [0, +\infty)$ such that:*

(i) h is continuous and V is of class C^1;
(ii) $h(0) = 0$, and $h(x) > 0$ for each $x \in \mathcal{A}$ ($x \neq 0$);
(iii) $V(0) = 0$, and $V(x) > 0$ for each $x \in \mathcal{A}$ ($x \neq 0$);
(iv') $\nabla V(x) \cdot f(x) \leq -h(x)$, for each $x \in \mathcal{A}$;
(v') $\lim_{k \to \infty} V(x_k) = +\infty$ for each sequence $\{x_k\}$ such that $x_k \in \mathcal{A}$ for $k = 1, 2, \ldots$, and either $x_k \to \bar{x} \in \partial \mathcal{A}$ or $\|x_k\| \to \infty$.

In [153], a function $V(x)$ is called a *maximal Liapunov function* if there exists a function h such that the pair V, h enjoys properties (i), (ii), (iii), (iv') and (v') listed in Theorem 2.12. Theorem 2.12 is proven in [153] under the additional assumption that f is globally Lipschitz continuous on $\mathcal{A}$. We remark that this assumption can be actually removed without affecting the conclusion, by virtue of the following Proposition 2.11. Indeed, it is clear that, $\varphi(x)$ being a strictly positive function of class C^1, replacing $f(x)$ by $\tilde{f}(x) = \varphi(x)f(x)$ does not affect inequality (iv'); moreover, the phase portrait of the system remains unchanged and so does the domain of attraction.

Proposition 2.11 *Let Ω be open and connected. Let $f : \Omega \to \mathbb{R}^n$ be a vector field of class C^1. Then there exists a strictly positive function $\varphi : \Omega \to (0, +\infty)$ of class C^1 such that the field $\tilde{f}(x) := \varphi(x)f(x)$ is Lipschitz continuous on Ω.*

Proof. Without loss of generality we may assume that $0 \in \Omega$. Let $\omega : \Omega \to \mathbb{R}^+$ denote a nonnegative function of class C^1 such that

$$\omega(x) = 0 \iff x = 0;$$
$$\omega(x) \to +\infty \iff x \to \partial\Omega \text{ or } \|x\| \to +\infty.$$

We seek φ in the form

$$\varphi(x) = \delta(\omega(x)),$$

where $\delta : \mathbb{R}^+ \to (0, +\infty)$ is a function of class C^1. We set $\tilde{f}(x) := \delta(\omega(x))\,f(x)$ and for all $r \in \mathbb{R}^+$

$$h(r) := 1 + \max\left\{\left|\frac{\partial f_i}{\partial x_j}(x)\right|;\ \left|\frac{\partial \omega}{\partial x_j}(x)\right| \cdot |f_i(x)|;\ x \in \Omega,\ \omega(x) \leq r;\ 1 \leq i,j \leq n\right\}$$

As

$$\frac{\partial \tilde{f}_i}{\partial x_j}(x) = \delta'(\omega(x))\frac{\partial \omega}{\partial x_j}(x)f_i(x) + \delta(\omega(x))\frac{\partial f_i}{\partial x_j}(x),$$

we only have to determine the function δ in such a way that both functions δh and $\delta' h$ are bounded on $\mathbb{R}^+$. Thus, we are done if we prove the following lemma.

Lemma 2.3 *Let $h : \mathbb{R}^+ \to [1, +\infty)$ be a continuous and nondecreasing function. Then there exist a constant $C > 0$ and a strictly positive function $\delta : \mathbb{R}^+ \to (0, +\infty)$ of class C^1 such that for all $r \geq 0$*

$$\delta(r)h(r) \leq C, \quad |\delta'(r)|h(r) \leq C.$$

It is clear that Lemma 2.3 is a direct consequence of the following result.

Lemma 2.4 *Let h be as in Lemma 2.3. Then there exists a strictly positive, nonincreasing and continuous function $\gamma : \mathbb{R}^+ \to (0, +\infty)$ such that for all $r \geq 0$*

$$\gamma(r)h(r) \leq 1, \quad \int_r^{+\infty} \gamma(s)\, ds \leq 3\gamma(r).$$

Indeed, setting $\delta(r) := \int_r^{+\infty} \gamma(s)\, ds$, we have $|\delta'(r)| = \gamma(r)$ and

$$\delta(r)h(r) \leq 3\gamma(r)h(r) \leq 3.$$

It remains to prove Lemma 2.4. We first define a sequence $(g_i)_{i \geq 0}$ of real numbers by induction on i. We set $g_0 = 1/h(0)$ and for each $i \geq 1$

$$g_i = \min\{\frac{1}{h(i)}, g_0\, e^{1-i}, g_1\, e^{2-i}, ..., g_{i-1}\}. \tag{2.21}$$

Notice that the sequence $g_0, g_1, ...$ is nonincreasing. Next, we define the linear piecewise function $g : \mathbb{R}^+ \to \mathbb{R}^+$ as follows: $g(i) = g_i$ for each $i \geq 0$ and g is linear on each interval $[i, i+1]$. We are now in a position to define the function γ. We set

$$\begin{aligned}
\gamma(r) &= \min\{\tfrac{1}{h(r)}, g(r)\} & \text{for } 0 \leq r < 1; \\
\gamma(r) &= \min\{\tfrac{1}{h(r)}, g(r), g_0\, e^{1-r}\} & \text{for } 1 \leq r < 2; \\
&\vdots \\
\gamma(r) &= \min\{\tfrac{1}{h(r)}, g(r), g_0\, e^{1-r}, g_1\, e^{2-r}, ..., g_{i-1}\, e^{i-r}\} & \text{for } i \leq r < i+1,
\end{aligned}$$

and so on. Using (2.21) we readily obtain $\gamma(i^+) = \gamma(i) = g_i$ and $\gamma(i^-) = g_i$. Therefore, the function γ is continuous. Clearly, γ is a nonincreasing positive function such that $\gamma(r)h(r) \leq 1$ for any $r \in \mathbb{R}^+$. Finally, pick any number $r \geq 0$ and let i denote the first integer such that $r \leq i - 1$. We write

$$\int_r^{+\infty} \gamma(s)\, ds = \int_r^i \gamma(s)\, ds + \int_i^{+\infty} \gamma(s)\, ds =: I_1 + I_2.$$

As γ is nonincreasing,

$$I_1 \leq (i - r)\gamma(r) \leq 2\gamma(r).$$

On the other hand, it follows from the definition of the function γ that $\gamma(s) \leq g_{i-1} e^{i-s}$ for all $s \geq i$. Therefore,

$$I_2 \leq g_{i-1} \int_i^{+\infty} e^{i-s}\, ds = g_{i-1} = \gamma(i-1) \leq \gamma(r).$$

We conclude that $\int_r^{+\infty} \gamma(s)\, ds = I_1 + I_2 \leq 3\gamma(r)$, as desired.

∎

As a nice illustration of Theorem 2.12, we report the following example (see again [153]).

Example 2.7 Let us consider the two dimensional system

$$\begin{cases} \dot{x}_1 = -x_1 + 2x_1^2 x_2 \\ \dot{x}_2 = -x_2 \, . \end{cases}$$

It is not difficult to find a strict Liapunov function in the small; for instance, with $V(x_1, x_2) = (x_1^2 + x_2^2)/2$ we obtain $\dot{V}(x_1, x_2) = -x_1^2(1 - 2x_1 x_2) - x_2^2$, which is negative definite in the region $\{(x_1, x_2) : x_1 x_2 < 1/2\}$. Hence, the origin is locally asymptotically stable. On the other hand, the domain of attraction is not the whole plane, since the solution $x_1 = e^t, x_2 = e^{-t}$ does not converge to the origin. The domain of attraction is actually given by the region $\mathcal{A} = \{(x_1, x_2) : x_1 x_2 < 1\}$ (some trajectories are plotted in the Figure 2.6). A maximal Liapunov function for this system has been computed in [153]: it is given by

$$V(x_1, x_2) = \frac{x_1^2 + x_2^2 - x_1 x_2^3}{2(1 - x_1 x_2)}\, .$$

∎

As already noticed, in order to find a maximal Liapunov function one should solve a partial differential equation, which in general is not easy. As an alternative, the following theorem can be used in order to approximate from inside the domain of attraction of system (2.13) by means of the level sets of any Liapunov function $V(x)$.

Theorem 2.13 *Let $\Omega \subseteq \mathbb{R}^n$ be an open set such that $0 \in \Omega$. Let $V : \Omega \to [0, +\infty)$ be such that:*

(1) $V(0) = 0$, and $V(x) > 0$ for each $x \in \Omega$, $x \neq 0$;

54   Time invariant systems

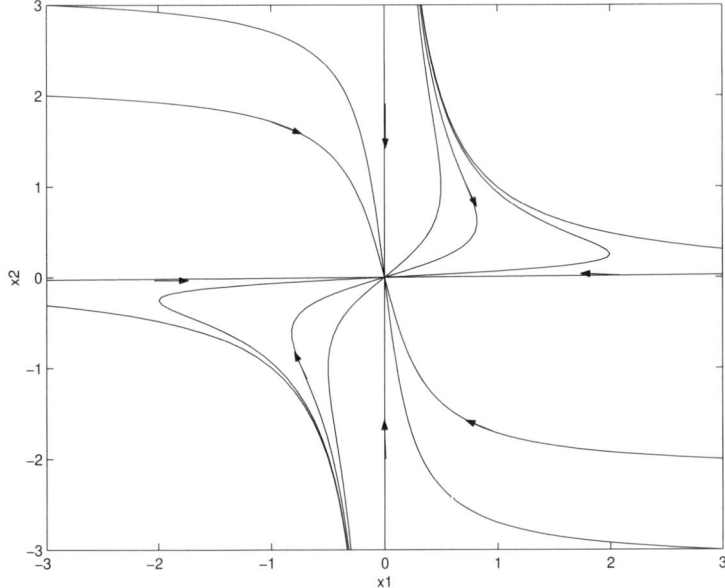

Figure 2.6: System of Example 2.6

(2) V is of class C^1;

(3) $\nabla V(x) \cdot f(x) < 0$, for each $x \in \Omega$, $x \neq 0$.

Let $l > 0$ and let Ω_l be the connected component of the level set $\{x \in \Omega : V(x) \leq l\}$ such that $0 \in \Omega_l$. Assume that Ω_l is a compact set. Then, the origin is asymptotically stable and $\Omega_l \subseteq \mathcal{A}$.

Proof The statement is actually a special instance of Theorem 2.10.

■

Example 2.8 The two dimensional system

$$\begin{cases} \dot{x}_1 = -x_1 + 2x_1^3 x_2^2 \\ \dot{x}_2 = -x_2 \end{cases}$$

is a classical example ([20], p. 68). A Liapunov function for this system can be written as

$$V(x_1, x_2) = \frac{x_1^2}{1 + x_1^2} + 2x_2^2 \ .$$

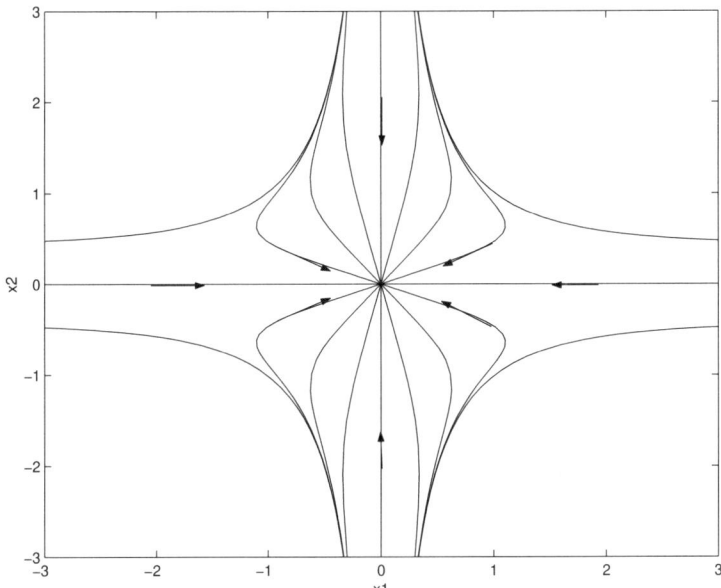

Figure 2.7: System of Example 2.7

Note that $\dot{V}(x_1, x_2)$ is strictly negative for all $(x_1, x_2) \in \mathbb{R}^2$; nevertheless, $V(x_1, x_2)$ is not radially unbounded, so we cannot conclude that the origin is globally asymptotically stable. As a matter of fact, even in this case we have a solution $x_1 = e^t, x_2 = e^{-t}$ which does not converge to the origin. Some trajectories of this system are shown in Figure 2.7. Figure 2.8 shows some level curves of the function $V(x_1, x_2)$. The domain of attraction is actually given by the region $\mathcal{A} = \{(x_1, x_2) : x_1^2 x_2^2 < 1\}$.

This example allows us to emphasize that the assumption about the compactness of Ω_l cannot be dropped in Theorem 2.13.

■

A completely different method for identifying the boundary of the domain of attraction, more geometric in nature and apparently not related with Liapunov method, is proposed in [33] (see also [34]).

2.2.8 Comparison functions

In order to formalize the general notions of external stability, we first introduce certain classes of functions to be used as comparison or "gain" functions.

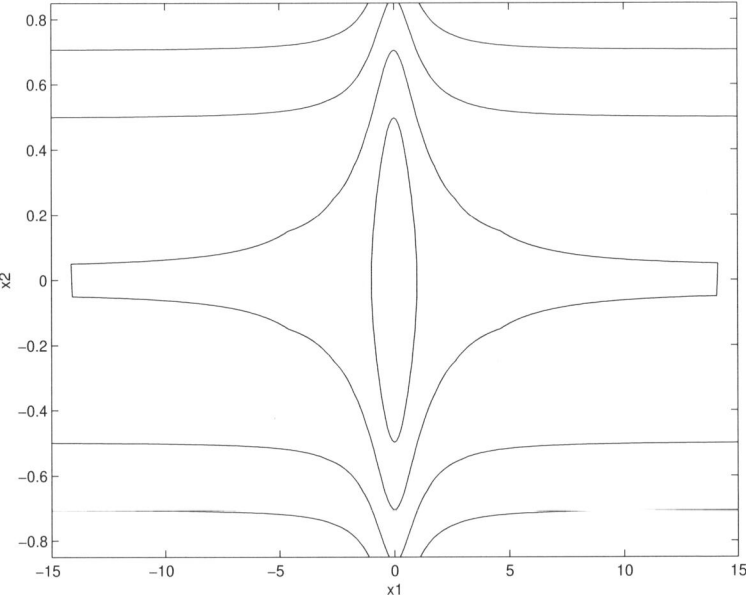

Figure 2.8: Level curves of the Liapunov function of Example 2.7

Comparison functions appear for the first time in the work of W. Hahn, who proposed for them the generic term "functions of class $\mathcal{K}$". In order to distinguish several types of such functions, we need to adopt here a more specific notation.

A function $\alpha : [0, r_1) \to [0, +\infty)$ is said to belong to the class $\mathcal{K}_0$ if it is continuous, strictly increasing and $\alpha(0) = 0$. Here, r_1 may be a positive number or $+\infty$ and may depend on α. When $\alpha \in \mathcal{K}_0$ and $r_1 = +\infty$, we say that α is of class $\overline{\mathcal{K}}_0$.

We say that a function $\alpha : [r_2, +\infty) \to [0, +\infty)$ (with $r_2 \geq 0$) is of class $\mathcal{K}^\infty$ if it is continuous, strictly increasing and $\lim_{r \to +\infty} \alpha(r) = +\infty$. If $\alpha \in \mathcal{K}^\infty$ and it is defined on $[0, +\infty)$, we write $\alpha \in \overline{\mathcal{K}}^\infty$

We also set $\mathcal{K}_0^\infty = \mathcal{K}_0 \cap \mathcal{K}^\infty = \overline{\mathcal{K}}_0 \cap \overline{\mathcal{K}}^\infty$.

A function $\alpha : [0, +\infty) \to [0, +\infty)$ is said to belong to the class $\mathcal{L}$ if it is continuous, decreasing and satisfies $\lim_{r \to +\infty} \alpha(r) = 0$.

A function $\beta : [0, +\infty) \times [0, +\infty) \to [0, +\infty)$ is said to belong to the class $\mathcal{LK}$ if it is of class $\mathcal{L}$ with respect to the first variable and of class $\overline{\mathcal{K}}_0$ with respect to the second one.

Comparison functions are an useful tool, which allows us to reformulate the definitions of stability in an elegant and unified way. To give an idea of how it works, we report here the following propositions, whose proofs can be found in [67], [3].

Proposition 2.12 *The following statements are equivalent:*

(i) system (2.13) is stable at the origin;

(ii) there exist r_0 $(0 < r_0 \leq +\infty)$ and a map $\alpha : [0, r_0) \to \mathbb{R}$, $\alpha \in \mathcal{K}_0$, such that for all x_0 with $\|x_0\| < r_0$, for all $x(\cdot) \in \mathcal{S}_{x_0}$ and for all $t \geq 0$

$$\|x(t)\| \leq \alpha(\|x_0\|) \ . \tag{2.22}$$

Proposition 2.13 *The following statements are equivalent:*

(i) the origin is globally asymptotically stable for (2.13)

(ii) there exists a map $\beta : [0, +\infty) \times [0, +\infty) \to \mathbb{R}^+$ which is of class $\mathcal{LK}$ and such that

$$\|x(t)\| \leq \beta(t, \|x_0\|) \tag{2.23}$$

for each $x_0 \in \mathbb{R}^n$, each $x(\cdot) \in \mathcal{S}_{x_0}$ and each $t \geq 0$.

Proposition 2.14 *The following statements are equivalent:*

(i) system (2.13) is Lagrange stable;

(ii) there exist r_1 $(0 \leq r_1 < +\infty)$ and a map $\alpha : [r_1, +\infty) \to \mathbb{R}$, $\alpha \in \mathcal{K}^\infty$, such that for all x_0 with $\|x_0\| > r_1$, for all $x(\cdot) \in \mathcal{S}_{x_0}$ and for all $t \geq 0$

$$\|x(t)\| \leq \alpha(\|x_0\|) \ . \tag{2.24}$$

In particular, Propositions 2.12 and 2.14 emphasize the analogy (and the difference) between Liapunov and Lagrange stability. In fact, Lagrange stability looks like Liapunov stability with respect to the "point at the infinity". Some authors consider Liapunov and Lagrange stability as dual properties.

Using comparison functions we can also express in a very useful way some typical features of Liapunov functions. Let $V(x)$ be a real function defined on B_r and such that $V(0) = 0$. If $V(x)$ is continuous and positive definite, then for $0 < r_0 < r$ we can find a function $a \in \mathcal{K}_0$ defined on $[0, r_0)$ such that

$$a(\|x\|) \leq V(x) \quad \text{for} \ \|x\| < r_0 \ . \tag{2.25}$$

Inequality (2.25) in turn implies that $V(x)$ is positive definite. Moreover, if V is defined on $\mathbb{R}^n$ and it is radially unbounded, then $a(\cdot)$ can be taken in the class $\mathcal{K}_0^\infty$.

We finally remark that if $V(x)$ is defined on B_r with $V(0) = 0$ $(0 < r \leq +\infty)$ and continuous, then there exists $b \in \mathcal{K}_0$ defined on $[0, r)$ such that

$$V(x) \leq b(\|x\|) \quad \text{for } \|x\| < r . \tag{2.26}$$

If in addition V is defined on $\mathbb{R}^n$ and radially unbounded, then $b \in \mathcal{K}_0^\infty$.

2.2.9 External notions

In this section we come back to the original model of a (time invariant) system with input

$$\dot{x} = f(x, u) \tag{2.27}$$

where f is a continuous map from $\mathbb{R}^n \times \mathbb{R}^m$ to $\mathbb{R}^n$. Note that for a given admissible input $u(t)$, the right hand side of (2.27) satisfies assumptions $(\mathbf{A}_1)$, $(\mathbf{A}_2)$ and $(\mathbf{A}_3)$ mentioned in Chapter 1. Therefore, the corresponding solutions should be intended in the Carathéodory sense.

We present now some extensions of the weak and strong finite gain properties previously encountered in the case of linear systems.

Definition 2.12 *We say that (2.27) is UBIBS-stable (i.e., uniformly bounded-input bounded-state stable) if for each $R > 0$, there exists $S > 0$:*

$$\|x_0\| \leq R , \ \|u(\cdot)\|_\infty \leq R \implies \|x(t)\| \leq S ,$$

$\forall t > 0$ *and for each solution* $x(\cdot) \in S_{x_0, u(\cdot)}$.

Equivalent definitions can be given in terms of comparison functions (see [3]).

Proposition 2.15 *The following statements are equivalent:*

(i) system (2.27) is UBIBS-stable;

(ii) there exist functions $\gamma_1, \gamma_2 \in \overline{\mathcal{K}}^\infty$ such that

$$\|x(t)\| \leq \gamma_1(\|x_0\|) + \gamma_2(\|u(\cdot)\|_\infty)$$

for each $t \geq 0$, each initial state x_0, each admissible input $u(\cdot)$ and each solution $x(\cdot) \in S_{x_0, u(\cdot)}$;

(iii) there exists $\Sigma \in \overline{\mathcal{K}}^\infty$ such that for each $R > 0$, each $x_0 \in \mathbb{R}^n$, each admissible input $u(\cdot)$ and each solution $x(\cdot) \in S_{x_0, u(\cdot)}$ one has

$$\|x_0\| \leq R, \quad \|u(\cdot)\|_\infty \leq R, \quad t \geq 0 \Longrightarrow \|x(t)\| \leq \Sigma(R) .$$

It is obvious that if (2.27) is UBIBS-stable, then it is internally Lagrange stable, but the converse is false in general.

The following notion has been introduced by E. Sontag ([131]).

Definition 2.13 *We say that system (2.27) possesses the* input-to-state stability *(in short, ISS) property, or that it is ISS-stable, if there exist maps $\beta \in \mathcal{LK}$, $\gamma \in \overline{\mathcal{K}}_0$ such that, for each initial state x_0, each admissible input $u : [0, +\infty) \to \mathbb{R}^m$, each solution $x(\cdot) \in \mathcal{S}_{x_0, u(\cdot)}$ and each $t \geq 0$*

$$\|x(t)\| \leq \beta(t, \|x_0\|) + \gamma(\|u\|_\infty) .$$

Note that if a system is ISS-stable, then the associated unforced system has a unique equilibrium position at the origin. The ISS stability property constrains the behavior of the system both for $\|x\|$ small and $\|x\|$ large. More precisely, it is not difficult to see that the ISS property implies that:

1) the system is internally globally asymptotically stable

2) the system is UBIBS-stable

On the contrary, the 1-dimensional example

$$\dot{x} = -(\cos u)^2 x$$

shows that a system may be internally globally asymptotically stable and UBIBS-stable but not ISS-stable.

Note also that the UBIBS stability property does not impose any constraint on the behavior of a system in a bounded region of the state space. In particular, a UBIBS-stable system may exhibit a large variety of nonlinear phenomena: multiple equilibrium positions, limit cycles, etc..

Even for UBIBS stability and the ISS property we can conceive appropriate Liapunov-like functions.

Definition 2.14 *A smooth* UBIBS-Liapunov function *for (2.27) is a radially unbounded C^1 function $V : \mathbb{R}^n \to \mathbb{R}$ which enjoys the following property: for each $R > 0$ there exists $N > 0$ such that*

$$\dot{V}(x, u) \stackrel{\text{def}}{=} \nabla V(x) \cdot f(x, u) \leq 0 \tag{2.28}$$

for each $x \in \mathbb{R}^n$ with $\|x\| \geq N$ and each $u \in \mathbb{R}^m$ with $\|u\| \leq R$.

Definition 2.15 *A smooth* ISS-Liapunov function *for (2.27) is a positive definite, radially unbounded C^1 function $V : \mathbb{R}^n \to \mathbb{R}$ with $V(0) = 0$ which enjoys further the following property: there exist functions $\rho, \chi \in \mathcal{K}_0^\infty$ such that for all $x \in \mathbb{R}^n$ ($x \neq 0$) and $u \in \mathbb{R}^m$, if $\|x\| \geq \rho(\|u\|)$ then*

$$\dot{V}(x,u) = \nabla V(x) \cdot f(x,u) < -\chi(\|x\|) \ . \tag{2.29}$$

The last inequality says that $\dot{V}(x,u)$ is negative definite uniformly with respect to u. The following result can be considered as an "external" version of Theorem 2.3 (the proof is originally given in [11], but see also the elder paper [154]).

Theorem 2.14 *Assume that system (2.27) possesses a smooth UBIBS-Liapunov function. Then, it is UBIBS-stable.*

Proof. Let R be any positive number, and let N be determined according to Definition 2.14. Without any loss of generality we can assume that $N > R$. Let $M = \max_{x \in \overline{B}_N} V(x)$, and let $S > N$ be such that $\|x\| > S \Longrightarrow V(x) > M$. Such a positive S exists, since V is radially unbounded. Now, let x_0 be such that $\|x_0\| \leq R$ and assume that for some admissible input $u(\cdot)$ with $\|u(t)\| \leq R$ a.e. for $t \in [0, +\infty)$, some $x(\cdot) \in \mathcal{S}_{x_0, u(\cdot)}$ and some $t_2 > 0$, one has

$$\|x(t_2)\| > S \ .$$

Since $x(t)$ is continuous, we can find $t_1 \in (0, t_2)$ such that $\|x(t_1)\| = N$ and $\|x(t)\| > N$ for $t \in (t_1, t_2]$. Thus, we have $V(x(t_1)) \leq M$ and $V(x(t_2)) > M$. On the other hand, the map $t \mapsto V(x(t))$ is differentiable a.e., and for a.e. $t \in [t_1, t_2]$

$$\frac{d}{dt} V(x(t)) = \nabla V(x(t)) \cdot f(x(t), u(t)) = \dot{V}(x(t), u(t)) \ .$$

By construction, (2.28) applies and we conclude that $V(x(t_1)) \geq V(x(t_2))$, a contradiction. ∎

A partial converse of this theorem has been recently given in [15]: under the assumption that (2.27) admits a unique solution for each admissible input, it is proven that UBIBS-stability implies the existence of a radially unbounded, upper semi-continuous function $V(x)$ which satisfies, instead of (2.28), the following monotonicity property:

$\forall R > 0, \exists N > 0$ *such that for each admissible input $u(\cdot)$, each solution $x(\cdot)$ defined on an interval I and corresponding to $u(\cdot)$, one has that the composite* map $t \mapsto V(x(t))$ is non-increasing on I, provided that $\|u(t)\| \leq R$ for a.e. $t \in I$ and $\|x(t)\| \geq N$ for each $t \in I$.

As the example at the end of Section 2.1.4 shows, the problem of the invertibility of Theorem 2.14 is far from being trivial even in the linear case.

We want now to present an "external" version of Theorem 2.2. It is essentially due to E. Sontag ([131]): we report a sketch of the proof for the sake of convenience.

Theorem 2.15 *Assume that there exists a ISS-Liapunov function for (2.27); then the system is ISS-stable.*

Proof. Since V is positive definite and radially unbounded, by (2.25), (2.26) we can write

$$a(\|x\|) \leq V(x) \leq b(\|x\|)$$

for each $x \in \mathbb{R}^n$, where a and b are some functions of class $\mathcal{K}_0^\infty$. Let $u(\cdot) : [0, +\infty) \to \mathbb{R}^m$ be a given measurable, essentially bounded input and let $x_0 \in \mathbb{R}^n$ be a given initial state. Let finally $x(t) \in \mathcal{S}_{x_0, u(\cdot)}$.

Let $c = b(\rho(\|u(\cdot)\|_\infty))$ and consider the compact set

$$K = \{x \in \mathbb{R}^n : V(x) \leq c\} .$$

Note that $x \in K$ implies $\|x\| \leq a^{-1}(b(\rho(\|u(\cdot)\|_\infty)))$, while $x \notin K$ implies $\|x\| > \rho(\|u(\cdot)\|_\infty)$. Moreover, K is positively invariant for $x(t)$, namely if for some t_0 it happens that $x(t_0) \in K$, than $x(t) \in K$ for each $t \geq t_0$. This can be proved by contradiction. Let $x(t_1) \notin K$ for some $t_1 > t_0$, which means $V(x(t_1)) > c$. Since $x(t)$ is continuous, there exists $\tau \in [t_0, t_1)$ such that $x(\tau) \in K$ and $x(t) \notin K$ for $t \in (\tau, t_1)$. In other words, $V(x(\tau)) \leq c < V(x(t_1))$. But for $x(t) \notin K$ we have

$$\|x(t)\| > \rho(\|u(\cdot)\|_\infty) \geq \rho(\|u(t)\|)$$

a.e.. Hence, condition (2.29) applies and $V(x(t))$ turns out to be decreasing. This yields $V(x(\tau)) > V(x(t_1))$, which is a contradiction.

We can now complete the argument. Recall that x_0 and the admissible input $u(\cdot)$ are given. If $x_0 \in K$, then setting $\gamma = a^{-1} \circ b \circ \rho$ we have $\|x(t)\| \leq \gamma(\|u(\cdot)\|_\infty)$ for each $t \geq 0$. If $x_0 \notin K$, the existence of a function $\beta \in \mathcal{LK}$ such that the inequality

$$\|x(t)\| \leq \beta(t, \|x_0\|)$$

holds as long as $x(t)$ remains outside K, can be proved by virtue of the comparison principle for solutions of differential equations (see for instance [70], p. 26) and Lemma 1 of [131]. In conclusion,

$$\|x(t)\| \leq \beta(t, \|x_0\|) + \gamma(\|u(\cdot)\|_\infty)$$

for each $t \in [0, +\infty)$. ∎

In fact, under the assumption that $f(x, u)$ is locally Lipschitz continuous, also the converse of Theorem 2.15 holds (see [142], [143], where many other characterizations of the ISS property are given). Theorem 2.15 can be reviewed as an extension of Proposition 2.7. Indeed, it is clear that a quadratic function $V(x) = x^t P x$ such that P satisfies the conditions of Proposition 2.7, is actually an ISS-Liapunov function for (2.1) in the sense of Definition 2.15.

2.3 Nonlinear systems: stabilization

As explained in the Introduction, and as already illustrated for the linear case, if the system is not stable (in some sense) we can try to achieve the desired property by the implementation of a feedback connection. Let us state first the basic definitions we are interested in.

Definition 2.16 *System (2.27) is said to be* continuously internally (locally or globally) stabilizable at the origin *if there exists a continuous feedback* $u = k(x)$ *such that the closed-loop system*

$$\dot{x} = f(x, k(x)) \tag{2.30}$$

is (locally or globally) asymptotically stable at the origin.

Note that if the system is stabilized by the feedback $u = k(x)$ in the sense of the previous definition, then $f(0, k(0)) = 0$ i.e., the origin is an equilibrium position for (2.30). The feedback $u = k(x)$ is also called an internal stabilizer.

Definition 2.17 *We say that (2.27) is* continuously ISS[UBIBS]-stabilizable *if there exists a continuous feedback* $u = k(x)$ *such that the closed-loop system*

$$\dot{x} = f(x, k(x) + v) \tag{2.31}$$

is ISS[UBIBS]-stable (here, v represents the external input).

Roughly speaking, a feedback $u = k(x)$ which provides ISS-stabilizability or UBIBS-stabilizability will be also called an external stabilizer. We need finally the following variant of the notion of Liapunov function (see [5], [132]).

Definition 2.18 *We say that (2.27) satisfies a smooth global control Liapunov condition (or that (2.27) has a smooth global control Liapunov function) if there exists a radially unbounded, positive definite, C^1 function $V(x)$ vanishing at the origin and enjoying the following property: for each $x \in \mathbb{R}^n \setminus \{0\}$ there exists $u \in \mathbb{R}^m$ such that*

$$\nabla V(x) \cdot f(x, u) < 0 . \qquad (2.32)$$

We have already encountered an example of control Liapunov function in the section concerning linear systems. Indeed, it is not difficult to recognize, that $V(x) = x^t P x$ is a global control Liapunov function for (2.1) if and only if (2.9) holds.

2.3.1 Necessary condition for internal stabilization

Definitions 2.16 and 2.17 above require the existence of a <u>continuous</u> feedback. This is consistent with the "smooth" setting of the present chapter. Note in particular that if $u = k(x)$ is continuous, then the right hand side of (2.30) is continuous, so that existence of solutions is guaranteed for each initial state.

In this section we focus our attention on Definition 2.16. Our purpose is to discuss the following natural question: how restrictive is it to insist on the continuity of the feedback law?

It is well known that if a linear system (2.1) can be internally stabilized by means of a continuous feedback $u = k(x)$, then it admits also a linear stabilizer $u = Fx$ (see [71]). In other words, stabilizability can be performed in such a way to preserve the form of the system. Unfortunately, this nice fact cannot be generalized. The next famous example (due to M. Kawski) shows that in general a nonlinear system cannot be stabilized by a feedback $u = k(x)$ which is at least as regular as the right hand side of (2.27) is.

Example 2.9 Consider the two-dimensional system

$$\begin{cases} \dot{x}_1 = u \\ \dot{x}_2 = x_2 - x_1^3 . \end{cases}$$

It cannot be stabilized by a feedback of class C^1, as a simple linearization argument shows, but there are explicit continuous locally asymptotically stabilizing feedback laws. One such a stabilizer is for instance

$$k(x) = -x_1 + x_2 + \frac{4}{3}x_2^{\frac{1}{3}} - x_1^3$$

(see [10] for details). ■

Similar examples can be given even if the state space is one-dimensional. For instance, the system

$$\dot{x} = x + 2u^3$$

can be stabilized by setting $u = -\sqrt[3]{x}$, but not by a feedback of class C^1. Note however that the right hand side is a nonlinear function of u.

The following examples aim to convince the reader that even the continuity requirement about the feedback law in Definitions 2.16 is restrictive.

Example 2.10 Let $r \geq 0$ be an arbitrary integer, and consider the one-dimensional example

$$\dot{x} = x^{2r+1} - 2ux^{2r}|x| \ . \tag{2.33}$$

Clearly, it is of class C^{2r}. The discontinuous feedback law $u = k(x) = \text{sgn}\, x$ renders the origin asymptotically stable, but if $k(x)$ is continuous, it is not difficult to see that there is an interval of the form $(0, \epsilon)$ or $(-\epsilon, 0)$ where the right hand side has the same sign than x. Hence, (2.33) is not continuously stabilizable (this example appears for the first time in [124], with $r = 0$).

Example 2.11 Another example is given by the one-dimensional equation

$$\dot{x} = x + ux^2 \ . \tag{2.34}$$

Whenever $u = k(x)$ is bounded, the linearization of the closed loop system presents a positive eigenvalue, hence it is unstable. However, the unbounded feedback $u = k(x) = -2/x$ gives rise to an asymptotically stable equation. ■

The non-existence of continuous internal stabilizers for (2.27) is related to certain obstructions of topological nature. The most famous one is pointed out by the following result, usually referred to as Brockett's test (see [26], [161], [10], [133]).

Theorem 2.16 *Consider the system (2.27) and assume that f is continuous and that $f(0,0) = 0$. A necessary condition for the existence of a continuous internal stabilizer $u = k(x)$ with $k(0) = 0$, is that for each $\varepsilon > 0$ there exists $\delta > 0$ such that*

$$\forall y \in B_\delta \; \exists x \in B_\varepsilon, \; \exists u \in B_\varepsilon \quad \text{such that} \quad y = f(x, u) \;.$$

In other words, f maps any neighborhood of the origin in $\mathbb{R}^{n+m}$ onto some neighborhood of the origin in $\mathbb{R}^n$. Brockett's test shows in particular that no system of the form

$$\dot{x} = \sum_{i=1}^{m} u_i f_i(x) \qquad (2.35)$$

where the vector fields f_i are continuous, $m < n$ and the linear subspace spanned by $f_1(0), \ldots, f_m(0)$ has dimension m, can be internally stabilized by a continuous feedback (the argument can be found in [136]).

Systems of the form (2.35) are sometimes called *nonholonomic*. Nonholonomic systems model a number of important applications (for instance in robotics). The prototype of a nonholonomic system is the so-called *nonholonomic integrator*

$$\begin{cases} \dot{x}_1 = u_1 \\ \dot{x}_2 = u_2 \\ \dot{x}_3 = x_1 u_2 - x_2 u_1 \;. \end{cases} \qquad (2.36)$$

Of course, Theorem 2.16 is not invertible. As an example, we can take equation (2.33): it is not difficult to see that it passes Brockett's test, and yet it is not continuously stabilizable. Another interesting example of this type[5] is given by the two-dimensional system

$$\begin{cases} \dot{x}_1 = u(x_1^2 - x_2^2) \\ \dot{x}_2 = 2u x_1 x_2 \end{cases} \qquad (2.37)$$

(see [136] for a thoughtful discussion; see also [138] and [109] for new interesting facts about this system).

All these examples motivate the interest in stabilizing feedback laws which are not necessarily continuous or time invariant: a first possible attempt is of course to relax the continuity requirement. Indeed, the question whether nonholonomic systems can be stabilized by means of discontinuous feedback has

---
[5] Example (2.37) comes back to [67], page 71; its reinterpretation in the framework of control theory is due to Z. Artstein ([5]), so that it is commonly known as Artstein's example.

been addressed by many authors. The answer is linked to the notion of solution adopted for the closed loop system which, if discontinuous feedback is allowed, turns out to have, of course, a discontinuous right hand side. Consider for instance system (2.37). It is clear that if solutions are intended in Carathéodory sense, then the feedback

$$u(x_1, x_2) = \begin{cases} -1 & \text{for } x_1 \geq 0 \\ 1 & \text{for } x_1 < 0 \end{cases} \tag{2.38}$$

is actually a stabilizer. However, as we shall see later (as a consequence of next Theorem 2.18 and Proposition 3.5), if the solutions are intended in Filippov's sense then no locally bounded feedback $u = k(x)$ can stabilize the system at the origin. As a matter of fact, with the feedback (2.38) all the points of the form $(0, x_2)$ are constant Filippov solutions for (2.37). A similar drawback arises with other feedback laws proposed in the literature (see for instance [29], [23]).

The following extension of Brockett's test is an easy consequence of a theorem proved by E. Ryan ([124]).

Theorem 2.17 *Consider a system of the form (2.27) and assume that the map $f(x, u) : \mathbb{R}^n \times \mathbb{R}^m \to \mathbb{R}^n$ is continuous, with $f(0, 0) = 0$. Assume in addition that for each subset $U \subset \mathbb{R}^m$ and each $x \in \mathbb{R}^n$,*

$$f(x, \mathrm{co}\, U) = \mathrm{co}\, f(x, U) . \tag{2.39}$$

Then, a necessary condition for the existence of a locally bounded, measurable feedback $u = k(x)$ which internally stabilizes (2.27) (in the sense of Filippov solutions) is that for each $\varepsilon > 0$ there exists $\delta > 0$ such that

$$\forall y \in B_\delta \ \exists x \in B_\varepsilon , \ \exists u \in \mathbb{R}^m \quad \text{such that} \quad y = f(x, u) .$$

Theorem 2.17 shows for instance that the nonholonomic integrator (2.36) cannot be stabilized (in Filippov's sense) by a locally bounded feedback $u = k(x)$. Another important aspect is the type of discontinuities which can be admitted for the feedback law. To this respect, note that the stabilizing feedback found for (2.33) is bounded, and the unique inevitable discontinuity is at the origin (the point to be stabilized). The stabilizing feedback found for (2.34) has again a unique discontinuity at the origin, but it is necessarily unbounded. We present a further example (basically due to E. Sontag, [136]). It shows that a system can admit a stabilizer which is continuous at the origin, but with inevitable discontinuities in any neighborhood of the origin.

Example 2.12 Consider a map $f(x,u) : \mathbb{R}^2 \to \mathbb{R}$ whose graph can be described in the following way. Let $\{A_i\}$ be a sequence of open strips of the plane (x, u) defined by

$$A_0 = \left\{(x,u) : 1 < x, \ \frac{3}{4} < u < 1\right\}$$

$$A_1 = \left\{(x,u) : \frac{1}{2} < x, \ \frac{3}{8} < u < \frac{1}{2}\right\}$$

$$\dots\dots\dots\dots\dots$$

$$A_i = \left\{(x,u) : \frac{1}{2^i} < x, \ \frac{1}{2^i} - \frac{1}{2^{i+2}} < u < \frac{1}{2^i}\right\}$$

$$\dots\dots\dots\dots\dots$$

and let $A = \cup A_i$. It is clear that the strips A_i "converge" to the positive x-axis and are pairwise nonoverlapping. Moreover, the projection of A on the positive x-axis is onto (see Figure 2.9).

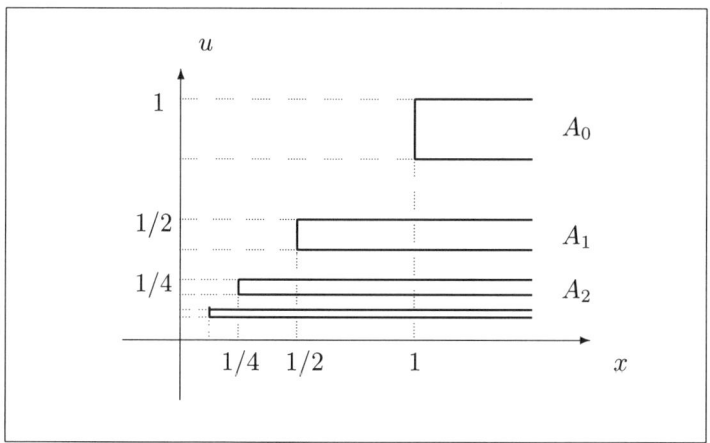

Figure 2.9: Strips where the map f of Example 2.11 is negative

The function $f(x, u)$ is assumed to be strictly negative on A and nonnegative outside A: in particular, strictly positive for $x < 0$, $u < 0$. In order to stabilize the system, we are forced to choose a function $u = k(x)$ in such a way that

$$f(x, k(x)) > 0 \ \text{ for } x < 0 \text{ and } f(x, k(x)) < 0 \text{ for } x > 0 \ .$$

While there is of course no problems with the first requirement, the second requirement can be met only if $u = k(x)$ has infinitely many discontinuities accumulating to the origin. ∎

Note that it makes sense to speak about Filippov solutions only if $u = k(x)$ is locally bounded.

We suggest to the reader the survey papers [136], [137] for a more exhaustive discussion about the need of introducing discontinuous feedback in stabilization theory. See also the recent paper ([31]).

2.3.2 Asymptotic controllability and local controllability

In this section we assume that $f(x, u)$ is continuous with respect to the pair $(x, u) \in \mathbb{R}^n \times \mathbb{R}^m$, and Lipschitz continuous with respect to x (uniformly with respect to u). We assume also that $f(0, 0) = 0$.

System (2.27) is said to be *globally asymptotically controllable* at the origin (see [37]) if there exist $C_0 > 0$, $C > 0$ such that:

(a) for each $x_0 \in \mathbb{R}^n$ there exists an admissible input $u_{x_0}(t) : [0, +\infty) \to \mathbb{R}^m$ such that the unique solution $x(t; 0, x_0, u_{x_0}(\cdot))$ is defined for all $t \geq 0$ and satisfies

$$\lim_{t \to +\infty} x(t; 0, x_0, u_{x_0}(\cdot)) = 0 \qquad (2.40)$$

(b) for each $\varepsilon > 0$ it is possible to find $\eta > 0$ such that if $\|x_0\| < \eta$ then there exists an admissible input $u_{x_0}(t)$ such that (2.40) holds, and in addition

$$\|x(t; 0, x_0, u_{x_0}(\cdot))\| < \varepsilon \quad \text{for all} \ \ t \geq 0 \qquad (2.41)$$

(c) if, in (b), the state x_0 satisfies also $\|x_0\| < C_0$, then the input $u_{x_0}(t)$ can be chosen in such a way

$$\|u_{x_0}(t)\| \leq C$$

for a.e. $t \geq 0$.

If (2.40) is required to hold only for each x_0 in some neighborhood of the origin, then we say that the system is *locally asymptotically controllable*.

The meaning of this definition is that the system is asymptotically driven toward zero by means of an open loop, bounded control which depends on the initial state.

Recall also that system (2.27) is said to be *small time locally controllable* (in short, STLC) to the origin if for each $T > 0$, there exists a $\delta > 0$ such that for each x_0 with $\|x_0\| < \delta$ there exists an admissible control $u_{x_0}: [0,T] \to \mathbb{R}^m$ such that

$$x(T; 0, x_0, u_{x_0}(\cdot)) = 0 \ . \tag{2.42}$$

As far as linear systems are concerned, STLC is equivalent to complete controllability. In the framework of geometric control theory, there is a wide literature about STLC (see [147] and the references therein). In particular, sufficient (and, separately, necessary) conditions for STLC can be found out by looking at the structure of the Lie algebra generated by the associated family of vector fields $\{X(x) = f(x,u): u \in \mathbb{R}^m\}$. Of course, these conditions apply only under the additional assumption that $x \mapsto f(x,u)$ is of class C^∞ for every $u \in \mathbb{R}^m$.

We are interested in some refinements of the STLC property. For instance, we shall say that (2.27) is STLC *with bounded controls* if there exists $C > 0$, and for each $T > 0$, there exists $\delta > 0$ such that for each x_0 with $\|x_0\| < \delta$ there exists an admissible control for which (2.42) holds and, in addition, $\|u_{x_0}(t)\| \leq C$ for a.e. $t \in [0,T]$.

We also say that (2.27) is STLC *with small controls* if for each $T > 0$, there exists $\delta > 0$ such that for each x_0 with $\|x_0\| < \delta$ there exists an admissible control for which (2.42) holds and, in addition, $\|u_{x_0}(t)\| \leq T$ for a.e. $t \in [0,T]$.

Roughly speaking, the difference between asymptotic controllability and small time local controllability is that in the first case the origin is "reached" asymptotically, while in the second case the origin in reached in finite time.

Proposition 2.16 *If the system (2.27) is STLC with bounded controls, then it is locally asymptotically controllable.*

Proof. Let C as in definition of STLC with bounded controls. Given $\varepsilon > 0$, let $M > \max\{\|f(x,u)\|, \text{ for } \|x\| \leq \varepsilon \text{ and } \|u\| \leq C\}$. Fix $T = \varepsilon/(3M)$ and let us find a corresponding δ according to the definition of STLC. We claim that (a), (b) and (c) hold with the same C and with $\eta < \min\{\varepsilon/3, \delta\}$.

In order to prove (a), it is sufficient to take a control $u_{x_0}(t)$ which steers x_0 to the origin at the instant T, and which vanishes for $t > T$. We know that this can be done with $\|u_{x_0}(t)\| \leq C$.

So, it remains to prove (b). Assume that for some $\bar{t} \in [0,T]$ we may have $\|x(\bar{t}; 0, x_0, u_{x_0}(\cdot))\| = \varepsilon$. Since the solution is continuous and $\eta < \varepsilon$ we may also assume that $\bar{t}$ is minimal i.e., $\|x(t; 0, x_0, u_{x_0}(\cdot))\| < \varepsilon$ for each $t \in [0, \bar{t})$. But

$$||x(\bar{t}; 0, x_0, u_{x_0}(\cdot))|| \leq ||x_0|| + \int_0^{\bar{t}} ||f(x(s; 0, x_0, u_{x_0}(\cdot)), u_{x_0}(s))|| \, ds$$
$$\leq \frac{\varepsilon}{3} + MT = \frac{2\varepsilon}{3}$$

which is a contradiction. ■

It is clear that if (2.27) is continuously internally stabilizable, then it is asymptotically controllable. The converse is true if the system is linear[6], but not in general. The classical counter-example is given by the nonholonomic integrator (2.36): it is STLC with bounded controls (it is a driftless system and its Lie algebra has a maximal rank at the origin, so that Chow's Theorem applies [80]) and hence asymptotically controllable: however, we know that it does not pass Brockett's test, so that it is not internally stabilizable.

Because of Ryan's extension of Brockett's test, it follows that large classes of asymptotically controllable systems can be stabilized not even by discontinuous feedback, at least as far as the solutions are intended in Filippov's sense. It has been recently proven in [37] that asymptotically controllable systems can be actually stabilized by time-sampled discontinuous feedback (see [2] and [111] for different solutions of the same problem).

2.3.3 Affine systems: internal stabilization

For the purposes of the next two subsections, we need to introduce some restrictions on the structure of the right hand side of (2.27). Recall that a system is said to be *affine with respect to the input* (or simply *affine*), when it has the form

$$\dot{x} = f_0(x) + \sum_{i=1}^{m} u_i f_i(x) \qquad (2.43)$$

where $f_0, f_1, \ldots, f_m$ are continuous vector fields of $\mathbb{R}^n$. There is a wide literature about internal stabilization of nonlinear (in particular, affine) systems. A relatively complete description of the state of the art at the end of 80's can be found in [134], [10]. However, in the last decade further very important progresses have been made in several directions, so that carrying out an updated survey seems to be today a very hard endeavor. It is in any case beyond the purposes of this book. Rather, we prefer to focus on a few results which fit

---

[6]In other words, for the linear case asymptotic controllability, STLC, stabilizability by continuous feedback and stabilizability by linear feedback are all equivalent: see [71].

in the thread of our argument, and which seem to be especially relevant for investigations about discontinuous feedback and Liapunov method.

When we deal with affine systems, it is customary to require $f_0(0) = 0$, and to look for feedback laws $u = k(x)$ which vanish at the origin [7].

We note also that affine systems meet the technical assumption (2.39) of Theorem 2.17.

Continuous stabilizers suffice

The first theorem below is due to Coron and Rosier ([49]). Essentially, it states that, as far as affine systems and Filippov solutions are concerned, the use of discontinuous feedback can be avoided.

Theorem 2.18 *Assume that system (2.43) can be internally stabilized (in the sense of Filippov solutions) at the origin by means of a locally bounded, measurable feedback $u = k(x)$ such that*

$$\lim_{\delta \to 0} \operatorname*{ess\,sup}_{\|x\| < \delta} \|u(x)\| = 0 \; . \tag{2.44}$$

Then, there exists also a continuous stabilizer $u = k(x)$ for the same system.

The assumption about the form of the system cannot be dropped out. This is shown by the following example.

Example 2.13 Consider the two-dimensional system

$$\begin{cases} \dot{x}_1 = u \\ \dot{x}_2 = x_1^2 - 2x_1 u^2 \; . \end{cases}$$

An explicit discontinuous stabilizing feedback (in the sense of Filippov solutions) fulfilling (2.44) is constructed in [49]. The system does not admit continuous stabilizers, since it does not pass Brockett's test (points of the form $(0, -\varepsilon)$ with $\varepsilon > 0$ do not belong to the image of $f(x, u)$). ∎

The role of control Liapunov functions

We want now to discuss possible generalizations of Proposition 2.6 to nonlinear systems. First of all, we need to add some new items to our terminology.

---

[7]It should be clear that this is actually a restriction. For instance, a simple linearization argument shows that the one-dimensional system $\dot{x} = x + ux$ does not admit continuous stabilizers such that $k(0) = 0$. Nevertheless, it can be stabilized setting $u = -2$.

A feedback law $u = k(x)$ is said to be *almost continuous* if it is continuous at every $x \in \mathbb{R}^n \setminus \{0\}$. Moreover, we say that (2.27) satisfies the *small control property* if for each $\varepsilon > 0$ there exists $\delta > 0$ such that for each $x \in B_\delta$, (2.32) is fulfilled for some $u \in B_\varepsilon$.

System (2.34) provides an example of a system which possesses a smooth control Liapunov function (take for instance $V(x) = x^2$) but not one fulfilling the small control property.

The following result is due to Z. Artstein ([5]; but see also [132])

Theorem 2.19 *If there exists a smooth global control Liapunov function for the affine system (2.43), then the system is globally internally stabilizable by an almost continuous feedback $u = k(x)$. If there exists a control Liapunov function which in addition satisfies the small control property, then it is possible to find a stabilizer $u = k(x)$ which is everywhere continuous.*

Vice-versa, if there exists a continuous global internal stabilizer $u = k(x)$, then there exists also a smooth global control Liapunov function and the small control property holds.

We do not report here the proof of this theorem, but some illustrative comments are appropriate. For the sake of simplicity, we limit ourselves to the single input case ($m = 1$). If the vector fields f_0 and f_1 are of class C^q ($0 \leq q \leq +\infty$) and a control Liapunov function of class C^r ($1 \leq r \leq +\infty$) is known, the stabilizing feedback whose existence is ensured by Theorem 2.19, can be explicitly constructed according to Sontag's "universal" formula

$$k(x) = \begin{cases} 0 & \text{if } b(x) = 0 \\ \frac{a(x) - \sqrt{a^2(x) + b^4(x)}}{b(x)} & \text{if } b(x) \neq 0 \end{cases} \quad (2.45)$$

where $a(x) = -\nabla V(x) \cdot f_0(x)$ and $b(x) = \nabla V(x) \cdot f_1(x)$ (see [132] for more details). We emphasize that such $k(x)$ is of class C^s (with $s = \min\{q, r-1\}$) on $\mathbb{R}^n \setminus \{0\}$. If the small control property is assumed, then the feedback law given by (2.45) turns out to be continuous also at the origin, but further regularity at the origin can be obtained only in very special situations.

It is worth noticing that the universal formula above has a powerful regularizing property. Indeed, if a continuous stabilizer for (2.43) is known, then Kurzweil's Converse Theorem applies. Hence, the existence of a C^∞ strict Liapunov function $V(x)$ for the closed loop system is guaranteed. It is not difficult to see that the same $V(x)$ is a control Liapunov function for (2.43). But then, the universal formula can be applied with this $V(x)$, and we obtain a new stabilizing feedback with the same order of differentiability as f_0 and f_1, (at least for $x \neq 0$). Feedback laws which are everywhere continuous and C^∞ for $x \neq 0$ are also called *almost smooth stabilizers*.

Theorem 2.19 does not hold for general systems of the form (2.27). This is shown for instance by the example (see [136]))

$$\begin{cases} \dot{x}_1 = u_2 u_3 \\ \dot{x}_2 = u_1 u_3 \\ \dot{x}_3 = u_1 u_2 \end{cases}$$

which possesses the control Liapunov function $V(x_1, x_2, x_3) = x_1^2 + x_2^2 + x_3^2$ but does not pass Brockett's test. However, the following extension holds (see the remark after Lemma 2.1 in [49]; see also [113]).

Theorem 2.20 *Consider a system of the form (2.27), where f is continuous and $f(0,0) = 0$. There exists a discontinuous feedback which stabilizes the system in Filippov's sense and which fulfills (2.44) if and only if there exists a smooth control Liapunov function for which the small control property holds.*

It is clear that if an affine system without drift (like the nonholonomic integrator (2.36) and the Artstein example (2.37)) admits a smooth control Liapunov function, then the small control property is automatically fulfilled. It follows from this simple remark and Theorem 2.19, that there exist no smooth control Liapunov functions for (2.36) and (2.37). Nevertheless, both systems are asymptotically controllable. This suggests the possibility of characterizing asymptotic controllability by some weaker notion of control Liapunov function.

Note that if the differentiability assumption about V is relaxed, then the monotonicity condition can be no more expressed in the form (2.32). In [130] (see also [134]) E. Sontag proved that if f is locally Lipschitz continuous with respect to both x, u, then the global asymptotic controllability is equivalent to the existence of a <u>continuous</u> global control Liapunov function. The monotonicity condition is expressed in [130] by means of Dini derivatives along the solutions (see Chapter 6). In [141] (see also [37]) it is pointed out that the same condition can be also expressed by means of contingent directional derivatives. Sontag's result has been recently improved ([111], [112]): every (globally) asymptotically controllable system with f locally Lipschitz continuous with respect to both x, u admits a <u>locally Lipschitz continuous</u> (global) control Liapunov function, provided that the monotonicity condition is expressed by means of the proximal gradient (see again Chapter 6). Note that this result applies in particular to Artstein's example (2.37).

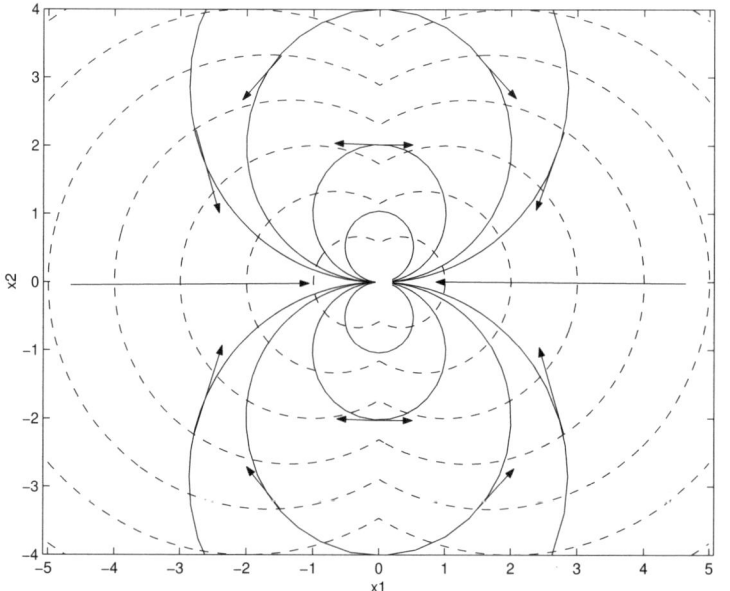

Figure 2.10: Artstein's example

Example 2.14 An explicit control Liapunov function for system (2.37) is given by

$$V(x_1, x_2) = \sqrt{4x_1^2 + 3x_2^2} - |x_1| . \tag{2.46}$$

It is clear that it is Lipschitz continuous. Moreover, for $x_1 > 0$, V is differentiable and we have

$$\nabla V(x_1, x_2) \cdot f(x_1, x_2, u) = u \left[\frac{x_1}{\sqrt{4x_1^2 + 3x_2^2}} \left(4x_1^2 + 2x_2^2 - x_1\sqrt{4x_1^2 + 3x_2^2} \right) + x_2^2 \right].$$

Since $4x_1^2 + 2x_2^2 > x_1\sqrt{4x_1^2 + 3x_2^2}$ for each $x_2 \in \mathbb{R}$ and each $x_1 > 0$, the expression in square brackets is always positive and (2.32) is fulfilled with $u = -1$. In a similar way, we see that for $x_1 < 0$, (2.32) is fulfilled with $u = 1$. The level curves of (2.46) are drawn in Figure 2.10 (dotted lines), together with the integral curves of (2.37) (solid lines). We see that (2.46) can be used as a nonsmooth Liapunov function and this explains why feedback (2.38) works.

For $x_1 = 0$, $x_2 \neq 0$, the Clarke gradient of V is given by $\{(p_1, p_2) : -1 \leq p_1 \leq 1, p_2 = (\operatorname{sgn} x_2)\sqrt{3}\}$. In particular, for $(p_1, p_2) = (0, \pm\sqrt{3})$, (2.32) is fulfilled for no value of u. On the contrary, (2.32) holds in the sense of proximal gradient also at points where $x_1 = 0$, since at these points the proximal gradient is empty.

A different locally Lipschitz continuous control Liapunov function for (2.37) is explicitly given in [137].

We emphasize that the tool used to express the monotonicity condition actually plays a crucial role. In [113] it is proved that if the monotonicity condition for a locally Lipschitz continuous control Liapunov function is expressed by means of Clarke gradient, then there exists also a smooth control Liapunov function, so that the system is stabilizable in Filippov sense.

Remark 2.7 It is possible to exhibit an explicit (global) locally Lipschitz Liapunov function for the holonomic integrator (2.36). Let us define

$$V(x_1, x_2, x_3) = \left|\sqrt{x_1^2 + x_2^2} - |x_3|\right| + 2|x_3| .$$

This is actually a control Liapunov function according to the definition of [130]. More precisely, for each nonzero initial state there exists a control $(u_1(t), u_2(t))$ such that the corresponding trajectory is driven to the origin, and the time derivative $\frac{d}{dt} V(x_1(t), x_2(t), x_3(t))$ is strictly negative (condition (2.4d) of [130]).

More precisely, let $r = \sqrt{x_1^2 + x_2^2}$, $\mathcal{U} = \{r > |x_3|\}$ and $\mathcal{O} = \{r < |x_3|\}$. Let $\bar{X} = (\bar{x}_1, \bar{x}_2, \bar{x}_3)$ be given, and let $\bar{r} = \sqrt{\bar{x}_1^2 + \bar{x}_2^2}$.

If $\bar{X} \in \mathcal{O}$ and $\bar{r} \neq 0$, we can take $(u_1, u_2) = (x_1, x_2)$, so that $dx_3/dt = 0$, $dr/dt = r$ and $dV/dt < 0$. After a finite time, we obtain $r = |x_3|$. If $\bar{r} = 0$, we take $(u_1, u_2) = (|\bar{x}_3|, 0)$ which implies $x_2 = 0$, $x_1 = t|\bar{x}_3|$, $x_3 = const$, $dV/dt < 0$.

If $\bar{X} \in \overline{\mathcal{U}}$ and $\bar{x}_3 \neq 0$, we take $(u_1, u_2) = (\operatorname{sgn}(\bar{x}_3) x_2, -\operatorname{sgn}(\bar{x}_3) x_1)$. It follows $dr/dt = 0$, $x_3 = \bar{x}_3 - (\operatorname{sgn} \bar{x}_3) t \bar{r}^2)$.

Finally, if $\bar{x}_3 = 0$, we take $(u_1, u_2) = (-\bar{x}_1, -\bar{x}_2)$.

Thus, for any initial state $\bar{X} \neq 0$, the origin is reached in finite time by using at most three different control laws and keeping $dV/dt < 0$. ∎

Damping control

Comparing the conclusions of Proposition 2.6 and Sontag's universal formula, we notice that there is a difference in the form of the feedback law. To this respect, it is interesting to ask under what assumptions a system can be stabilized by a feedback law of the simple form

$$u_i = -\alpha \nabla V(x) \cdot f_i(x) \qquad (2.47)$$

which is the natural extension of (2.10). Feedback laws like (2.47) are also called *damping control*. They have an important role to play in nonlinear stabilization.

Assume that the origin is a stable equilibrium for the unforced system associated to (2.43), and assume that a C^1 Liapunov function $V(x)$ such that $\nabla V(x) \cdot f_0(x) \leq 0$ for each x has been found. The set $\{x: \nabla V(x) \cdot f_0(x) = 0\}$ will be called the *bad set*. In a similar manner, for any feedback law $u = k(x)$ we can consider the bad set of the closed loop system (again with respect to V).

Lemma 2.5 *Assume that a weak Liapunov function of class C^1 for the unforced system associated to (2.43) is known, and let $u = k(x)$ be given by (2.47). Then, for any $\alpha > 0$, the bad set of the closed loop system is contained in the bad set of the unforced system associated to (2.43).*

The meaning of Lemma 2.5 is that the stability performances of the closed loop system are in general better (or at least are not worse) than those of the unforced associated system.

Lemma 2.5 is a crucial step of the so-called Jurdjevic-Quinn method for smooth stabilization (see [81], [61], [56] and [10] for related developments). In fact, using the Invariance Principle and an additional Lie algebraic condition, Jurdjevic and Quinn were able to prove that the feedback law (2.47) actually stabilizes the system. More precisely, assume that the vector fields appearing in (2.43) are C^∞. Recall that the Lie bracket operator associates to an (ordered) pair g_0, g_1 of vector fields the vector field

$$[g_0, g_1] = Dg_1 \cdot g_0 - Dg_0 \cdot g_1$$

(here, Dg_i denotes the Jacobian matrix of g_i, $i = 0, 1$). The "ad" operator is iteratively defined by

$$ad_{g_0}^1 g_1 = [g_0, g_1] \qquad ad_{g_0}^{k+1} g_1 = [g_0, ad_{g_0}^k g_1] \,.$$

Theorem 2.21 (JURDJEVIC-QUINN) *Assume that a weak Liapunov function of class C^∞ for the unforced system associated to (2.43) is known, and let $u = k(x)$ be given by (2.47). Assume further that for $x \neq 0$ in a neighborhood of the origin*

$$\dim \text{span}\,\{f_0(x), ad^k_{f_0}f_i(x),\ i=1,\ldots m,\ k=1,2,\ldots\} = n\ .$$

Then, for any $\alpha > 0$, the system is stabilized by (2.47).

A final comment about Theorem 2.21 is appropriate. According to what has been already seen in Subsection 2.2.2, when the unforced system has a Liapunov stable equilibrium position at the origin the existence of a smooth weak Liapunov function in the small cannot be given for sure. If we have at our disposal a Liapunov function which is at least locally Lipschitz continuous, then (2.47) can be defined almost everywhere. It can be proved that if $f_0(x)$ is continuous, then the conclusion of Lemma 2.5 remains valid ([12]). Note that in these conditions, in general (2.47) turns out to be discontinuous.

2.3.4 Affine systems: external stabilization

The claim that the external behavior is determined by the internal one is no more valid in the nonlinear case. This is shown for instance by the following simple example

$$\dot{x} = -x + ux^3\ . \tag{2.48}$$

The unforced system is globally asymptotically stable but with $u \equiv 1$ there are unbounded solutions. However, some connections can be recovered provided that the use of feedback is allowed. The first step on our way is the following theorem, due to E. Sontag ([131]).

Theorem 2.22 *Assume that a smooth global strict Liapunov function $V(x)$ for the unforced system associated to (2.43) is known. Then, there exists a continuous feedback $u_i = k_i(x)$ ($i = 1,\ldots,m$) such that the same $V(x)$ is a smooth ISS-Liapunov function for the closed loop system*

$$\dot{x} = f_0(x) + \sum_{i=1}^{m}[k_i(x) + v_i]f_i(x) \tag{2.49}$$

where $v = (v_1,\ldots,v_m)$ represents the external input.

The proof of the previous theorem is constructive. Focusing again for simplicity on the single input case, one can take for instance $k(x) = -a(x)\cdot b(x)/2$ where, as before, $a(x) = -\nabla V(x) \cdot f_0(x)$ and $b(x) = \nabla V(x) \cdot f_1(x)$ (see [131] for more details). Note that under the assumption of Theorem 2.22 and by virtue of Theorem 2.15, the closed loop system (2.49) turns out to be ISS-stable.

Note also that if (2.43) is internally globally asymptotically stabilizable by a continuous feedback, then Kurzweil's Theorem guarantees the existence of a global strict Liapunov function $V(x)$ of class C^∞ for the (unforced) closed loop system, so that a ISS stabilizer can be found with the same regularity as f_0 and f_1.

In conclusion, we arrive at the following important result (Sontag, [131]).

Theorem 2.23 *Every internally globally stable (or continuously internally globally stabilizable) affine system of the form (2.43) is ISS-stabilizable.*

Next, we want to discuss the possibility of repeating, as far as possible, the same reasoning concerning the UBIBS property. Looking at Proposition 2.5, and recalling that in the linear case internal (Liapunov) stability is equivalent to Lagrange stability, it seems natural to identify Lagrange stability as the right internal property to be compared with UBIBS-stability.

Next theorem is an analogous of Theorem 2.22 for UBIBS-stability (for a proof, we refer to [11]).

Theorem 2.24 *Assume that a smooth weak Liapunov function in the large $V(x)$ for the unforced system associated to (2.43) is known. Then, there exists a feedback law $u_i = k_i(x)$ ($i = 1, \ldots, m$) such that the same $V(x)$ is a UBIBS-Liapunov function for the closed loop system (2.49).*

In the single input case, the feedback law mentioned in Theorem 2.24 can be taken, for instance, of the form

$$k(x) = -||x|| \operatorname{sgn}(\nabla V(x) \cdot f_1(x)) .$$

It must be emphasized that this expression is not continuous in general. Fortunately, Theorem 2.14 remains valid even if the right hand side of the system possesses discontinuities with respect to x and the solutions are intended in Filippov's sense.

To complete the picture, one would need a converse theorem for Lagrange stability. However, we already know that it is possible to construct semicontinuous Liapunov functions for Lagrange stable systems, but further regularity cannot be guaranteed, in general. In conclusion, the best we can do is the following.

Theorem 2.25 *Assume that the unforced system associated to (2.43) is Lagrange stable and that it admits a smooth weak Liapunov function in the large. Then, it is UBIBS-stabilizable (in the sense of Filippov solutions).*

2.4 Output systems

An accurate representation of a physical system usually involves a large number of state variables. However, not all of them can be observed. Moreover, in general it is not required to control all these variables, but only a few of them or a combination of them. This leads to consider, together with system (2.27), an observation map

$$y = h(x) \tag{2.50}$$

where $h : \mathbb{R}^n \to \mathbb{R}^p$ is at least continuous. The variable $y \in \mathbb{R}^p$ (where p is a given positive integer) is called the *output*.

Several notions introduced in the previous sections can be reformulated in terms of the output. It is worth noticing that the method of Liapunov functions can be usefully employed also in this setting. A detailed survey on this subject is out of the aim of this book (we refer the readers to the original research papers [131], [135], [144], [145], [146]). However, for the sake of completeness, we recall the following generalization of the ISS property ([144]).

Definition 2.19 *System (2.27) is said to be* input-to-output stable *(in short, IOS-stable) if the following two properties hold:*

(i) There exist two maps $\sigma_1, \sigma_2 \in \mathcal{K}_0^\infty$ such that for each admissible input $u(\cdot)$, each initial state $x_0 \in \mathbb{R}^n$, each solution $x(\cdot) \in \mathcal{S}_{x_0, u(\cdot)}$ and each $t \geq 0$, one has

$$\|x(t)\| \leq \sigma_1(\|x_0\|) + \sigma_2(\|u\|_\infty)$$

(ii) there exist $\gamma \in \overline{\mathcal{K}}_0$ and $\beta \in \mathcal{LK}$ such that for each admissible input $u(\cdot)$, each initial state $x_0 \in \mathbb{R}^n$, each solution $x(\cdot) \subset \mathcal{S}_{x_0, u(\cdot)}$ and each $t \geq 0$, one has

$$\|h(x(t))\| \leq \beta(t, \|x_0\|) + \gamma(\|u\|_\infty) \ .$$

The property (i) above is a strengthened version of the UBIBS property, since the gain functions σ_1, σ_2 are required to be of class $\mathcal{K}_0^\infty$. As the UBIBS property, it expresses the fact that for bounded input, the whole state vector remains bounded: but, in addition, it imposes that if the initial state and the input are small, the whole state vector remains small. The second property (ii) has the same meaning as the ISS property, but it is referred to the output, instead of the state vector.

In [131], it is proved that IOS and ISS stability are equivalent, provided that the system satisfies suitable controllability and observability conditions. The characterization of the IOS property in terms of Liapunov functions is given in [146].

2.5 Cascade systems

In the class of single-input affine systems, special interest is deserved in systems of the form

$$\begin{cases} \dot{y} = f(y, z) \\ \dot{z} = u \end{cases} \quad (2.51)$$

where $x = (y, z) \in \mathbb{R}^n$ $(n \geq 2)$, $y \in \mathbb{R}^{n-1}$ and $z \in \mathbb{R}$. They can be reviewed as a description of a plant whose state vector is y and whose input z is generated by an integrator. For this reason, (2.51) is usually called a *cascade connection* or a *cascade system*. Throughout this section, we always assume for simplicity that $f(0, 0) = 0$.

The recent paper [30] provides another interesting motivation for systems of the form (2.51). To describe it, consider a single-input, affine system

$$\dot{x} = f_0(x) + u f_1(x) \quad (2.52)$$

where f_0 and f_1 are analytic vector fields of $\mathbb{R}^n$.

Roughly speaking, and according to the differential geometry tradition, a map which associates to each point of $\mathbb{R}^n$ a subspace of $\mathbb{R}^n$ will be called a *distribution*. Recall that a distribution Δ is *involutive* if for each pair of vector fields f, g, the conditions $f(x) \in \Delta(x)$ and $g(x) \in \Delta(x)$ for each x imply $[f, g](x) \in \Delta(x)$.

Special distributions related to (2.52) can be defined by setting

$$\Delta_0(x) = \operatorname{span}\{f_1(x)\}$$
$$\Delta_1(x) = \operatorname{span}\{f_1(x), ad_{f_0}^1 f_1(x)\}$$
$$\Delta_{n-1}(x) = \operatorname{span}\{f_1(x), ad_{f_0}^1 f_1(x), \ldots, ad_{f_0}^{n-1} f_1(x)\}.$$

The following result is proved in [30].

Proposition 2.17 *Assume that (2.52) is STLC, and assume further that there exists an open set U such that $\overline{U}$ is a neighborhood of the origin and the distributions $\Delta_0(x), \ldots, \Delta_{n-1}(x)$ are involutive and have constant dimensions for*

$x \in U$. Then, up to a local change of coordinates, (2.52) can be put in the so-called triangular form, that is

$$\begin{cases} \dot{x}_1 = f_{0,1}(x_1, x_2) \\ \dot{x}_2 = f_{0,2}(x_1, x_2, x_3) \\ \quad \vdots \\ \dot{x}_{n-1} = f_{0,n-1}(x_1, \ldots, x_n) \\ \dot{x}_n = f_{0,n}(x_1, \ldots, x_n) + u f_{1,n}(x_1, \ldots, x_n) \end{cases} \quad (2.53)$$

with $f_{1,n}(0) \neq 0$.

Note that by possibly applying a further local change of coordinates and a feedback transformation, (2.53) becomes a particular case of (2.51). The class of systems identified by the assumptions of Proposition 2.17 includes the class of affine systems which are linearizable by feedback equivalence.

Another reason of interest in systems of the form (2.51) is the following result, due to M. Kawski ([85]).

Theorem 2.26 *Assume that $n = 2$ and that the map f is real analytic. Assume also that system (2.51) is STLC. Then, it is locally stabilizable at the origin by means of a continuous feedback.*

Unfortunately, under these assumptions we cannot hope to find in general differentiable feedback laws: this point has been already discussed in Section 2.3.1 (see in particular Example 2.9).

We consider now the following problem. Assume that the plant

$$\dot{y} = f(y, z) \quad (2.54)$$

is stabilizable at the origin by means of a feedback law $z = h(y)$. Does there exist a feedback of the form $u = k(y, z)$ stabilizing at the origin the cascade system (2.51)? In other words, is it possible to inject a stabilizing feedback into the plant (2.54) through a cascade connection? This problem is also known as *stabilization under addition of an integrator*. We report the following theorem, due to J. Tsinias ([149], but see also [27], [140]).

Theorem 2.27 *Assume that in (2.51), f is of class C^r ($r \geq 1$). If there exists a map $z = h(y)$ of class C^r such that the system*

$$\dot{y} = f(y, h(y)) \quad (2.55)$$

is locally asymptotically stable at the origin, then the overall system (2.51) is locally stabilizable at the origin by means of a feedback law of class C^{r-1}.

Theorem 2.27 suggests the idea that stabilization by adding an integrator is possible in general, but at the cost of a lack of regularity. Nevertheless, we know several examples which seem to support the opposite intuition.

Example 2.15 It is not difficult to see that the one-dimensional equation

$$\dot{y} = y^3 - z^9$$

can be stabilized by a continuous feedback (setting for instance $z = \sqrt[3]{2y}$) but not by a differentiable one. On the other hand, the cascade system

$$\begin{cases} \dot{y} = y^3 - z^9 \\ \dot{z} = u \end{cases}$$

can be stabilized by the polynomial feedback $u = 2y - 3y^2 z + y^3 - z^9$. The argument is based on the local Liapunov function $V(y, z) = y^2(1-yz) + z^{10}/10$ and the Invariance Principle ([10], [53]). ∎

This example shows that at least under certain circumstances, the dynamical extension introduced by adding an integrator enables us to improve the regularity of the feedback law.

An interesting question, not answered by Theorem 2.27, is what happens when the $(n-1)$-dimensional plant (2.54) can be stabilized by means of a continuous feedback $u = h(y)$, but not by a differentiable one. It turns out that the answer depends on the value of n and the type of discontinuities allowed for the feedback law. The proof of the following proposition can be found in [150].

Proposition 2.18 *Let n be any integer, $n \geq 2$. Assume that the map $f(y,z)$ is continuous. Assume in addition that the plant (2.54) is locally stabilizable at the origin by a continuous feedback $u = h(y)$. Then, the cascade system (2.51) is locally stabilizable by a feedback law $u = k(y,z)$ which may be discontinuous and unbounded at the origin.*

The following result (see [117, Chapter VI]) is useful in situations where the $(n-1)$-dimensional plant (2.54) can be stabilized at the origin by a continuous feedback $u = h(y)$, for which the gradient $\nabla h(y)$ exists for $y \neq 0$ and satisfies the growth condition

$$||\nabla h(y)|| = o(||y||^{-1}) \text{ as } y \to 0.$$

Proposition 2.19 *Let n be any integer ($n \geq 2$) and assume that the map $f(y,z)$ is continuous. Assume in addition that there exists a feedback law $u = h(y)$, $h \in C^1(\Omega \setminus \{0\}, \mathbb{R}) \cap C^0(\Omega, \mathbb{R})$ (Ω is some open neighborhood of $0 \in \mathbb{R}^{n-1}$) stabilizing locally the plant (2.54) at the origin, and fulfilling*

$$\sup_{\{z;\ 0 \leq h(y)z \leq h(y)^2\}} |\nabla h(y) \cdot f(y,z)| \to 0 \text{ as } y \to 0. \tag{2.56}$$

Then, the cascade system (2.51) is locally stabilizable by a continuous feedback law $u = k(y,z)$.

The proof of Proposition 2.19 is reported in the Appendix of this chapter. For two-dimensional cascade systems, we have the following result (see [53]).

Proposition 2.20 *Let $n = 2$ and assume that $f(y,z)$ is real analytic in a neighborhood of the origin. Assume further that there exists a feedback law $z = h(y)$ (possibly discontinuous) such that the one-dimensional closed loop system (2.55) is locally asymptotically stable at the origin in the sense of Filippov solutions. Then, system (2.51) is locally stabilizable at the origin by means of a continuous feedback.*

If the right hand side of the plant $f(y,z)$ is assumed to be only continuous instead of analytic, then it may happen that the plant is stabilizable by a continuous feedback, but the cascade system (2.51) is stabilizable not even by a discontinuous feedback which satisfies condition (2.44) (see [49] for an example). In [49] we can find also an example of a real analytic system with $n = 3$ (i.e., the cascade connection of a two-dimensional plant and an integrator) such the plant is stabilizable by a discontinuous feedback which satisfies condition (2.44), but the overall system is not stabilizable by a discontinuous feedback which satisfies condition (2.44) (here again, stabilization is intended in the sense of Filippov solutions).

2.6 Appendix

Before giving the proof of Proposition 2.19, we say a few words about the condition (2.56). Along any trajectory $y(t)$ of (2.55) we have

$$\frac{d\,h(y)}{dt} = \nabla h(y) \cdot f(y, h(y)).$$

Since $h(y(t)) \to 0$ as $t \to +\infty$, it is quite natural to expect that $d\,h(y)/dt \to 0$, too. Therefore, the assumption

$$\lim_{y \to 0} \nabla h(y) \cdot f(y, h(y)) = 0 \tag{2.57}$$

seems not to be too restrictive in practice. The assumption (2.56) is just a strengthened version of (2.57). The following lemma gives a natural framework in which Proposition 2.19 may be applied.

Lemma 2.6 *Let the functions $f(y,z)$ and $h(y)$ fulfill the following properties:*
(i) $\lim_{y \to 0} ||y|| \cdot ||\nabla h(y)|| = 0;$
(ii) $\exists C > 0, \; ||f(y, h(y))|| \leq C||y|| \quad \forall y \in \Omega;$
(iii) *f is Lipschitz continuous in a neighborhood of the origin.*
Then the condition (2.56) is satisfied.

Proof. Let $y \in \Omega$ and let z be any number between 0 and $h(y)$. According to the intermediate value theorem, there exists some point $\bar{y}$ in the segment $[0, y]$ such that $h(\bar{y}) = z$. If L denotes some Lipschitz constant for f in a neighborhood of the origin, we may write

$$\begin{aligned}
|\nabla h(y) \cdot f(y,z)| &= |\nabla h(y) \cdot (f(y,z) - f(\bar{y},z)) + \nabla h(y) \cdot f(\bar{y}, h(\bar{y}))| \\
&\leq ||\nabla h(y)|| \cdot ||f(y,z) - f(\bar{y},z)|| + ||\nabla h(y)|| \cdot ||f(\bar{y}, h(\bar{y}))|| \\
&\leq (L||y - \bar{y}|| + C||\bar{y}||) \cdot ||\nabla h(y)|| \\
&\leq (L + C)||y|| \cdot ||\nabla h(y)||,
\end{aligned}$$

as $||y - \bar{y}|| \leq ||y||$. Then, letting $y \to 0$ and using (i), we obtain (2.56). ∎

At this point it is worth noticing that the conclusion of Proposition 2.19 is no longer true if we replace (2.56) by (2.57), or (i)-(ii) (see [117]). We now turn to the proof of Proposition 2.19. We aim to design a *piecewise continuous* feedback law $k(y,z)$ satisfying the condition

$$k(y,z) \to 0 \text{ as } ||(y,z)|| \to 0,$$

and stabilizing the system

$$\begin{cases} \dot{y} = f(y,z), \\ \dot{z} = k(y,z) \end{cases} \quad (2.58)$$

at the origin, the solutions being taken in the Filippov sense. As (2.58) is affine in the control this guarantees, according to [49, Theorem 1.5], the existence of a *continuous* feedback $\tilde{k}(y,z)$ law stabilizing (2.58) at the origin. The discontinuous feedback $k(y,z)$ is designed in such a way that the manifold $\{z = h(y)\}$ is positively invariant by the flow of (2.58), and it is reached in finite time by any trajectory issued from an initial position close to the origin. This implies that on the invariant manifold

$$\dot{z} = \frac{d\,h(y)}{dt} = \nabla h(y) \cdot f(y, h(y)).$$

We now define the piecewise continuous feedback law $k(y, z)$ in the following way. If $h(y) \geq 0$, we set

$$k(y, z) = -(z - h(y))^{\frac{1}{3}} + \begin{cases} -(|\nabla h(y)| + 1) \cdot |f(y, z)| & \text{if } h(y) < z; \\ |\nabla h(y) \cdot f(y, z)| & \text{if } 0 < z < h(y); \\ |f(y, z)| & \text{if } z < 0, \end{cases}$$

and if $h(y) < 0$, we set

$$k(y, z) = -(z - h(y))^{\frac{1}{3}} + \begin{cases} -|f(y, z)| & \text{if } 0 < z; \\ -|\nabla h(y) \cdot f(y, z)| & \text{if } h(y) < z < 0; \\ (|\nabla h(y)| + 1) \cdot |f(y, z)| & \text{if } z < h(y). \end{cases}$$

Recall that the value of the function k on the zero measure set $\{z = 0\} \cup \{z = h(y)\}$ plays no role in the construction of the Filippov solutions to (2.58). Let

$$G = \{(y, z) \in \Omega \times \mathbb{R}; \ 0 < z < h(y) \ \text{ or } \ h(y) < z < 0\}.$$

Let $x_0 = (y_0, z_0) \in \Omega \times \mathbb{R}$, and let $x(t) = (y(t), z(t))$ denote a Filippov solution of (2.58) issuing from x_0 at $t = 0$.

CLAIM 1. If $x_0 \notin \overline{G}$, then the positive orbit $\{x(t); \ t \geq 0\}$ reaches ∂G in finite time.

Indeed, it follows from the definition of the function $k(y, z)$ that

$$||\dot{z}|| \geq ||\dot{y}|| + |z - h(y)|^{\frac{1}{3}}, \qquad (2.59)$$

hence there exists some time $T > 0$ such that $z(T) = 0$ or $z(T) = h(y(T))$. Integrating in (2.59) we obtain

$$|z(t) - z_0| \geq ||y(t) - y_0|| \quad 0 \leq t \leq T,$$

which, when combined to the fact that the function $|z(t)|$ is nonincreasing, yields

$$||y(t) - y_0|| + |z(t)| \leq |z_0| \quad 0 \leq t \leq T.$$

Therefore the stability property is fulfilled up to the time T when the trajectory reaches ∂G.

CLAIM 2. The manifold $M_1 := \{(y, z) \in \Omega \times \mathbb{R}; \ z = h(y)\}$ is invariant. Assume e.g. that $h(y) < 0$. As the number $\nabla h(y) \cdot f(y, h(y))$ belongs to the segment

$$[-|\nabla h(y) \cdot f(y, h(y))|, (|\nabla h(y)| + 1) \cdot |f(y, h(y))|] =: K_y \, k(y, h(y)),$$

we infer that for any trajectory $y(t)$ of (2.55), the map $x(t) = (y(t), h(y(t)))$ is a Filippov solution of (2.58). Furthermore, as

$$\frac{d|z - h(y)|}{dt} = \operatorname{sgn}(z - h(y))(\dot{z} - \nabla h(y) \cdot \dot{y}) \leq -|z - h(y)|^{\frac{1}{3}}$$

we see that every trajectory touching M_1 stays actually on M_1 in the future. The situation is quite different for the submanifold $M_2 = \{(y, 0); h(y) \neq 0\}$. Indeed, the trajectory issued from an initial point (y_0, z_0) satisfying $z_0 \leq 0 < h(y_0)$, after reaching M_2, goes into G, since

$$\dot{z} \geq -(z - h(y))^{\frac{1}{3}} > 0.$$

By symmetry, the same situation occurs for a trajectory issued from an initial point (y_0, z_0) satisfying $h(y_0) < 0 \leq z_0$.

It remains to deal with the case where $x_0 \in G$. Let us consider a small number $\alpha > 0$ such that

$$\|f(y, z)\| < 1 \text{ if } \|(y, z)\| < \alpha \qquad (2.60)$$

and

$$B_\alpha(0) \subset \Omega. \qquad (2.61)$$

(In what follows, we denote by B a ball in $\mathbb{R}^n$, and by $\mathcal{B}$ a ball in $\mathbb{R}^{n+1}$.) Let $D := B_\alpha((0,0)) \cap G$, and let $x(\cdot) = (y(\cdot), z(\cdot)) : I \to \overline{D}$ denote a maximal solution (in the Filippov sense) to the Cauchy problem

$$\begin{cases} \dot{y} = f(y, z) & (y, z) \in \overline{D} \\ \dot{z} = k(y, z) \\ (y(0), z(0)) = (y_0, z_0) \in D. \end{cases}$$

(We stress that the property $(y(t), z(t)) \in \overline{D}$ is required for all $t \in I$.) Set $I^+ := I \cap \mathbb{R}^+$. As long as $(y, z) \in \overline{D}$ we have

$$\frac{d|z - h(y)|}{dt} \leq -|z - h(y)|^{\frac{1}{3}}.$$

Recall that the unique solution to the Cauchy problem

$$\begin{cases} \dot{v} = -v^{\frac{1}{3}} \\ v(0) = v_0 > 0 \end{cases}$$

is given by

$$v(t) = \begin{cases} (v_0^{\frac{2}{3}} - \frac{2}{3}t)^{\frac{3}{2}} & \text{if } 0 \leq t \leq \frac{3}{2}v_0^{\frac{2}{3}}, \\ 0 & \text{if } t \geq \frac{3}{2}v_0^{\frac{2}{3}}. \end{cases}$$

Setting
$$T_{y_0,z_0} = \frac{3}{2}(z_0 - h(y_0))^{\frac{2}{3}},$$
we conclude that if $T_{y_0,z_0} \in I^+$ then for all $t \geq T_{y_0,z_0}$, $z(t) - h(y(t)) = 0$ and $\dot{y}(t) = f(y,h(y))$, hence $I^+ = \mathbb{R}^+$, $y(t) \to 0$ and $z(t) = h(y(t)) \to 0$ as $t \to +\infty$, provided that $y(T_{y_0,z_0})$ is small enough. To complete the proof we need to justify the following

CLAIM 3. For (y_0, z_0) close enough to 0, $T_{y_0,z_0} \in I^+$ and $\|y(T_{y_0,z_0})\| < \alpha$.

Pick a number $\varepsilon > 0$ such that
$$B_\varepsilon(0) \times (-\varepsilon, \varepsilon) \subset \mathcal{B}_\alpha((0,0)),$$
and a number $R < \varepsilon$ such that every maximal solution of (2.55) issuing from the ball $B_R(0)$ at $t = 0$ is defined on $\mathbb{R}^+$, stays within $B_\varepsilon(0)$, and tends to 0 as $t \to +\infty$. The feedback h being continuous at the origin, we may define a function $r(\delta)$ such that $0 < r(\delta) < R$ for each δ, $r(\delta) \to 0$ as $\delta \to 0$, and
$$\|y\| < r(\delta) \Rightarrow |h(y)| < \delta.$$

Pick some number $\delta_1 \in (0, \varepsilon/2)$, and write $r_1 = r(\delta_1)$. If $\|y_0\| < r_1$ and $|z_0| < \delta_1$, then $|z_0 - h(y_0)| < 2\delta_1$, hence
$$T_{y_0,z_0} = \frac{3}{2}(z_0 - h(y_0))^{\frac{2}{3}} \leq \frac{3}{2}(2\delta_1)^{\frac{2}{3}}.$$

Using (2.60), we get for any $t \in I \cap [0, T_{y_0,z_0}]$
$$\|y(t)\| \leq \|y_0\| + \int_0^t \|f(y(s), z(s))\|\, ds \leq r_1 + T_{y_0,z_0} \leq r_1 + \frac{3}{2}(2\delta_1)^{\frac{2}{3}}.$$

Let now $\delta_2 > 0$ be such that $\delta_2 < \min(\delta_1, \varepsilon/4)$ and $r_2 + \frac{3}{2}(2\delta_2)^{\frac{2}{3}} < r_1$, where r_2 denotes the number $r(\delta_2)$. According to the previous computation (applied to (δ_2, r_2) instead of (δ_1, r_1)), we obtain for $\|y_0\| < r_2$ and $|z_0| < \delta_2$
$$\|y(t)\| < r_1 \text{ and } |h(y(t))| < \delta_1 \quad \forall t \in [0, T_{y_0,z_0}] \cap I.$$

As
$$|z(t) - h(y(t))| \leq |z_0 - h(y_0)| < 2\delta_2 < \frac{\varepsilon}{2},$$
we infer that
$$(y(t), z(t)) \in K := \overline{B}_{r_1}(0) \times [-(\delta_1 + \frac{\varepsilon}{2}), \delta_1 + \frac{\varepsilon}{2}].$$
As $\delta_1 < \varepsilon/2$, we readily see that K is a compact set contained in the ball $\mathcal{B}_\alpha((0,0))$. It follows that $T_{y_0,z_0} \in I$ and $\|y(T_{y_0,z_0})\| < r_1 < R$. Therefore,

we conclude that $I^+ = \mathbb{R}^+$, $||y(t)|| < \varepsilon$ for all $t \geq 0$ (distinguish the two cases $t \leq T_{y_0,z_0}$ and $t \geq T_{y_0,z_0}$),

$$|z(t)| \leq \begin{cases} \varepsilon & \text{if } t \leq T_{y_0,z_0} \\ \sup_{B_\varepsilon(0)} |h(\cdot)| & \text{if } t \geq T_{y_0,z_0}, \end{cases}$$

and that $(y(t), z(t)) \to (0,0)$ as $t \to +\infty$. The proofs of Claim 3 and of Proposition 2.19 are completed. ■

Chapter 3

Time varying systems

The discussion of the previous chapter shows that there are at least two problems of interest in nonlinear control theory, namely the existence of Liapunov functions for Liapunov stable or Lagrange stable systems and the existence of internal stabilizers, which in general do not have a solution in the class of C^1 or even continuous functions.

Semi-continuous Liapunov functions are easier to construct, but then, it becomes more difficult to check that they are non-increasing along the trajectories. To this purpose, we can try to make use of nonsmooth analysis methods: this aspect of the problem will be considered later in Chapter 6.

As far as the stabilization problem is concerned, we know examples of systems which are stabilized by discontinuous feedback, but not by a continuous one. Moreover, there are also classes of systems (for instance, nonholonomic systems), that can't even be stabilized by a discontinuous feedback.

In this chapter we explore an alternative way to address the aforementioned problems. We ask whether more regularity can be achieved, provided that solutions are sought in some larger classes of functions. More precisely, from now on we shall admit Liapunov functions and stabilizing feedback laws which depend explicitly on time, that is of the form $V(t,x)$ or, respectively, $u = k(t,x)$.

3.1 Two examples

The aim of next two examples is to show that, by removing the time-invariance constraint, we actually have better chances of success.

Example 3.1 Consider the scalar equation

$$\dot{x} = f(x) \tag{3.1}$$

where $f(x) = x \cos^2(\log |x|)$ for $x \neq 0$, and $f(0) = 0$. It is clear that $f(x)$ is continuous at the origin, and analytic for $x \neq 0$. Moreover, the origin is stable, since it is a limit point of two sequences of equilibria

$$\left\{\exp\left(\frac{\pi}{2} + k\pi\right)\right\}_{k \in \mathbb{Z}}, \quad \left\{-\exp\left(\frac{\pi}{2} + k\pi\right)\right\}_{k \in \mathbb{Z}}.$$

Actually, equation (3.1) exhibits the same features as Example 2.2 in Chapter 2. In particular, we know that there exists no time invariant Liapunov function which is continuous in a whole neighborhood of the origin. Nevertheless, we claim that there exists a function $V(t, x)$ which is positive, everywhere continuous and, at the same time, non-increasing along the solutions of (3.1).

The construction of such a function can be accomplished according to an idea of Yoshizawa ([160]; see also the following Theorem 3.5). Let $t_0 \geq 0$ and $x_0 \in \mathbb{R}$ be given, and define

$$V(t_0, x_0) = \min_{0 \leq s \leq t_0} |x(s)| \tag{3.2}$$

where $x(t)$ denotes the (unique) solution of (3.1) such that $x(t_0) = x_0$. In order to prove that (3.2) fulfills all the mentioned requirements, it would be desirable to have a more explicit expression. This is possible, by virtue of the particular choice of $f(x)$. We can limit ourselves to $x > 0$ (since f is odd, we clearly have $V(t, -x) = V(t, x)$). Provided that x_0 is not an equilibrium of (3.1), by separation of variables we obtain

$$\int_{x_0}^{x(s)} \frac{1}{x \cos^2(\log x)} \, dx = \int_{t_0}^{s} dt = s - t_0 .$$

Using the change of variable $y = \log x$, we have

$$\int_{x_0}^{x(s)} \frac{1}{x \cos^2(\log x)} \, dx = \int_{\log x_0}^{\log x(s)} \frac{1}{\cos^2 y} \, dy = \tan(\log x(s)) - \tan(\log x_0) .$$

Hence,

$$\tan(\log x(s)) = \tan(\log x_0) + s - t_0 . \tag{3.3}$$

Let us set for simplicity

$$k_0 = \left[\frac{\log x_0}{\pi} + \frac{1}{2}\right]$$

where the square brackets $[\,\cdot\,]$ denote the integer part. Since x_0 is not an equilibrium, we have

$$k_0\pi - \frac{\pi}{2} < \log x_0 < k_0\pi + \frac{\pi}{2} \,. \qquad (3.4)$$

In fact, because of the uniqueness of solutions, the same inequalities are fulfilled by $\log x(s)$, for each $s \in \mathbb{R}$. From (3.4) it follows

$$\log x_0 = k_0\pi + \arctan(\tan(\log x_0)) \qquad (3.5)$$

and, from (3.3),

$$\log x(s) = k_0\pi + \arctan(s - t_0 + \tan(\log x_0)) \,.$$

Finally,

$$x(s) = \exp(k_0\pi + \arctan(s - t_0 + \tan(\log x_0))) \,.$$

This is an increasing function of s. Hence, substituting in (3.2), we get

$$V(t_0, x_0) = \exp\left(k_0\pi + \arctan(-t_0 + \tan(\log x_0))\right) \,.$$

Summing up, for $t \geq 0$ and $x \geq 0$ the values of $V(t,x)$ are determined according to the following rule:

$$V(t,x) = \begin{cases} 0 & \text{if } x = 0 \\ x & \text{if } x = \exp(\frac{\pi}{2} + k\pi), \ k \in \mathbb{Z} \\ \exp\left(\left[\frac{\log x}{\pi} + \frac{1}{2}\right]\pi \right. & \\ \left. + \arctan(-t + \tan(\log x))\right) & \text{otherwise} \,. \end{cases}$$

Next, we verify that $V(t,x)$ possesses all the required properties. By direct computation, it is not difficult to see that $\frac{d}{dt}V(t,x(t))$ exists and vanishes, for every solution $x(t)$ of (3.1). It is obvious that V is positive: in fact, from (3.4), it follows that

$$\exp(\log x - \pi) \leq V(t,x) \,. \qquad (3.6)$$

In other words, V is positive uniformly with respect to t. From (3.2), it is also clear that

$$V(t, x) \leq x \ . \tag{3.7}$$

This implies in particular that V is continuous at every point of the form $(t_0, 0)$. Continuity at the points of the form (t_0, x_0), where x_0 is an equilibrium of (3.1), can be directly verified by using the explicit expression of V. If x_0 is not an equilibrium, continuity of V at (t_0, x_0) is obvious.

Example 3.2 This example is taken from [106]. It concerns a three-dimensional time invariant system with two independent inputs which cannot be stabilized at the origin by means of a time invariant continuous feedback. Nevertheless, a continuous time dependent (actually periodic) stabilizing feedback can be explicitly constructed. The system is

$$\begin{cases} \dot{x} = u \\ \dot{y} = xv \\ \dot{z} = v \end{cases} \tag{3.8}$$

where u, v represent the inputs. It is clear that the right hand side does not meet Brockett's test (no vector of the form $(0, \varepsilon, 0)$ belongs to its image). According to the method developed in [106], let us define

$$u = y \sin t - (x + y \cos t) \quad \text{and} \quad v = -x(x + y \cos t) \cos t - (xy + z) \ .$$

In order to verify that this is actually a stabilizer, we introduce the following (time varying) Liapunov function

$$V(t, x, y, z) = \frac{(x + y \cos t)^2}{2} + \frac{y^2 + z^2}{2} \ .$$

Note that

$$u = y \sin t - \frac{\partial V}{\partial x}(t, x, y, z) \quad \text{and} \quad v = -x \frac{\partial V}{\partial y}(t, x, y, z) - \frac{\partial V}{\partial z}(t, x, y, z) \ .$$

Let us denote by $\dot{V}(t, x, y, z)$ the function

$$\frac{\partial V}{\partial t}(t, x, y, z) + \frac{\partial V}{\partial x}(t, x, y, z)u + \frac{\partial V}{\partial y}(t, x, y, z)xv + \frac{\partial V}{\partial z}(t, x, y, z)v \ .$$

A simple computation yields

$$\begin{aligned}\dot{\tilde V}(t,x,y,z) &= -y(x+y\cos t)\sin t + \frac{\partial V}{\partial x}(t,x,y,z)\big[y\sin t - \frac{\partial V}{\partial x}(t,x,y,z)\big]\\ &\quad + x\frac{\partial V}{\partial y}(t,x,y,z)\big[-x\frac{\partial V}{\partial y}(t,x,y,z) - \frac{\partial V}{\partial z}(t,x,y,z)\big]\\ &\quad + \frac{\partial V}{\partial z}(t,x,y,z)\big[-x\frac{\partial V}{\partial y}(t,x,y,z) - \frac{\partial V}{\partial z}(t,x,y,z)\big]\\ &= -\big[\frac{\partial V}{\partial x}(t,x,y,z)\big]^2 - \big[x\frac{\partial V}{\partial y}(t,x,y,z) + \frac{\partial V}{\partial z}(t,x,y,z)\big]^2\\ &= -[x+y\cos t]^2 - [x(x+y\cos t)\cos t + xy + z]^2\ .\end{aligned}$$

Now it is easily recognized that this last expression is negative semi-definite, but actually not negative definite. More precisely, we have

$$\dot{\tilde V}(t,x,y,z) = 0 \iff \begin{cases} x+y\cos t = 0\\ xy+z = 0\ .\end{cases} \qquad (3.9)$$

According to the periodic time version of the LaSalle invariance principle (see [121], p. 50), we can conclude that the origin is asymptotically stable provided that two further conditions are fulfilled. The first condition requires that no positive trajectory (except the trivial one) is contained in the set where $\dot{\tilde V}$ vanishes. This condition is actually satisfied in our case. Indeed, on this set the closed loop system reduces to

$$\begin{cases} \dot x = y\sin t\\ \dot y = 0\\ \dot z = 0 \end{cases}$$

whose solutions are of the form $x = a - b\cos t$, $y = b$ and $z = c$. It is not difficult to check that no such curve (except the case $a = b = c = 0$) lies on (3.9).

The second condition requires that $V(t,x,y,z) \geq \alpha(||(x,y,z)||)$ for some function of class $\mathcal{K}_0$. To this end, we observe that for any fixed point $(x,y,z) \neq 0$, $m(x,y,z) = \min_t V(t,x,y,z)$ is strictly positive and continuous, as a function of (x,y,z). Hence, for each $r > 0$ we can find $h > 0$ and a positive number $\tilde m(r)$ such that

$$\tilde m(r) \leq V(t,x,y,z)$$

for each t, and each (x,y,z) such that $r-h \leq ||(x,y,z)|| \leq r+h$. The required function α can now be constructed by standard technical arguments.

Remark 3.1 It is well known that the change of coordinates

$$x_1 = z, \quad x_2 = x, \quad x_3 = xz - 2y$$

transforms (3.8) into (2.36) (with $u_1 = v$, $u_2 = u$). Thus, we see that the nonholonomic integrator (2.36) can be globally stabilized by means of a time varying feedback

$$\begin{aligned} u_1 &= -x_2^2 \cos t - \frac{x_2}{2}(x_1 x_2 - x_3)(\cos^2 t + 1) - x_1 \\ u_2 &= \frac{x_1 x_2 - x_3}{2}(\sin t - \cos t) - x_2 \ . \end{aligned}$$

3.2 Reformulation of the basic definitions

The extension of the stability notions and the definitions of Liapunov function in a time varying context is not straightforward at all. Indeed, new delicate aspects emerge, especially related to uniformity with respect to time. Of course, there are many good books devoted to this topic (among them, we recall [67], [121]). However, for reader's convenience we collect here the main definitions and a short discussion about their relationships. We also report some interesting counter-examples which are essentially well known but not easy to locate in the classical literature, and we finally present a complete "map" of the situation.

3.2.1 Stability and attraction

In this section we are mainly interested in the characterization of the internal behavior of a time varying system. We know that this is the same as studying classical stability properties of a system without input. In view of the main purposes of this book, a system of this type should be always thought of as the unforced system associated to some system of the form (1). However, for simplicity, throughout the present and the next five sections we adopt the shortened notation

$$\dot{x} = f(t, x) \tag{3.10}$$

with $x \in \mathbb{R}^n$ and $t \in [0, +\infty)$. According to the notation introduced in Chapter 1, for any given pair of initial conditions $(t_0, x_0) \in [0, +\infty) \times \mathbb{R}^n$, the set of all local Carathéodory solutions $x(\cdot)$ of (3.10) such that $x(t_0) = x_0$ is denoted by $\mathcal{S}_{t_0,x_0}$. Moreover, in order to guarantee that $\mathcal{S}_{t_0,x_0}$ is nonempty for each pair (t_0, x_0), we limit ourselves to systems whose right hand side $f(t,x)$ satisfies at least conditions ($\mathbf{A}_1$), ($\mathbf{A}_2$) and ($\mathbf{A}_3$) (see again Chapter 1). This has to be considered as a standing assumption throughout the present section.

The origin is said to be an *equilibrium position* for (3.10) if $f(t,0) = 0$ for a.e. $t \geq 0$.

Definition 3.1 *We say that (3.10) is stable at the origin (or that the origin is stable for (3.10)) if for each $t_0 \geq 0$ and for each $\varepsilon > 0$ there exists $\delta > 0$ such that for each $||x_0|| < \delta$ and each solution $x(\cdot) \in \mathcal{S}_{t_0,x_0}$ one has that $x(\cdot)$ is continuable on $[t_0, +\infty)$ and*

$$||x(t)|| < \varepsilon \quad \text{for each } t \geq t_0 \; .$$

Note that in Definition 3.1 existence of solutions is not strictly necessary from an abstract point of view. However, existence of solutions is a quite natural requirement, especially in view of applications. Note also that if the origin is stable then for each t_0, $x(t) \equiv 0$ is the unique solution such that $x(t_0) = 0$. Hence, the origin is an equilibrium position.

In Definition 3.1, δ may depend on both ε and t_0.

Definition 3.2 *We say that (3.10) is uniformly stable at the origin (or that the origin is uniformly stable for (3.10)) if for each $\varepsilon > 0$ there exists $\delta > 0$ such that for each $t_0 \geq 0$, each $||x_0|| < \delta$ and each solution $x(\cdot) \in \mathcal{S}_{t_0,x_0}$ one has that $x(\cdot)$ is continuable on $[t_0, +\infty)$ and*

$$||x(t)|| < \varepsilon \quad \text{for each } t \geq t_0 \; .$$

Definition 3.3 *System (3.10) is said to be locally attractive at the origin if for each $t_0 \geq 0$ there exists $\delta_0 > 0$ such that for each x_0 with $||x_0|| \leq \delta_0$ and each $x(\cdot) \in \mathcal{S}_{t_0,x_0}$, one has*

$$\lim_{t \to +\infty} ||x(t)|| = 0 \; .$$

Note that the origin may be attractive without being an equilibrium point. As an example, we can take the scalar equation

$$\dot{x} = -x + e^{-t} \cos t$$

(incidentally, we observe that the origin cannot be stable in this case).

Again, we can conceive more specific notions by requiring that the limit in Definition 3.3 is uniform with respect to the initial state, or with respect to both the initial state and the initial time.

Definition 3.4 *We say that system (3.10) is* locally equi-attractive *at the origin (or that the origin is locally equi-attractive for (3.10)) if for each $t_0 \geq 0$ there exists $\delta_0 > 0$ such that for each $\sigma > 0$ there exists $T = T(\sigma, t_0) > 0$ such that for each $\|x_0\| < \delta_0$, and each solution $x(\cdot) \in \mathcal{S}_{t_0, x_0}$, one has that $x(\cdot)$ is continuable on $[t_0, +\infty)$ and*

$$\|x(t)\| < \sigma \quad \text{for each } t \geq t_0 + T.$$

Definition 3.5 *We say that system (3.10) is* uniformly locally attractive *at the origin (or that the origin is uniformly locally attractive for (3.10)) if there exists $\delta_0 > 0$ such that for each $\sigma > 0$ there exists $T = T(\sigma) > 0$ such that for each $\|x_0\| < \delta_0$, each $t_0 \geq 0$, and each solution $x(\cdot) \in \mathcal{S}_{t_0, x_0}$ one has that $x(\cdot)$ is continuable on $[t_0, +\infty)$ and*

$$\|x(t)\| < \sigma \quad \text{for each } t \geq t_0 + T. \tag{3.11}$$

The origin is uniformly globally attractive *if for each $\sigma > 0$ there exists $T = T(\sigma) > 0$ such that (3.11) holds for each $x_0 \in \mathbb{R}^n$, each $t_0 \geq 0$, each $t \geq t_0 + T$, and each solution $x(\cdot) \in \mathcal{S}_{t_0, x_0}$.*

The definitions above can be combined in several ways. We are in particular interested in the following properties.

Definition 3.6 *We say that system (3.10) is* locally [globally] asymptotically stable *at the origin (or that the origin is locally [globally] asymptotically stable for (3.10)) if (3.10) is stable and locally [globally] attractive.*

We say that system (3.10) is uniformly locally [globally] asymptotically stable *at the origin (or that the origin is uniformly locally [globally] asymptotically stable for (3.10)) if (3.10) is uniformly stable and uniformly locally [globally] attractive.*

It is evident that uniform stability implies stability. Moreover, uniform attraction implies equi-attraction which in turn implies attraction. The situation further simplifies for those systems which satisfy a local Lipschitz continuity condition.

Let us recall in particular that under assumptions ($\mathbf{A}_1$), ..., ($\mathbf{A}_4$), solutions are locally unique and the property of continuity with respect to the initial data holds (property (C) of Chapter 1).

Proposition 3.1 *Assume that the origin is an equilibrium position, and assume that* $(\mathbf{A}_1)$, ..., $(\mathbf{A}_4)$ *are satisfied for (3.10). Then, equi-attraction implies stability.*

Proof. Let t_0 and $\varepsilon > 0$ be fixed, and apply Definition 3.4 with $\sigma = \varepsilon$. We find δ_0 and $T(\varepsilon, t_0)$ such that the inequality

$$||x(t; t_0, x_0)|| < \varepsilon \qquad (3.12)$$

holds for $t > t_0 + T(\varepsilon, t_0)$, provided that $||x_0|| < \delta_0$. On the other hand, applying (C) with $\bar{t} = t_0$ and $\bar{x} = 0$, we can find $\delta = \delta(\varepsilon, t_0)$ such that if $||x_0|| < \delta$ then (3.12) holds for $t \in [t_0, t_0 + T(\varepsilon, t_0)]$. The statement follows, since it is not restrictive to assume $\delta < \delta_0$. ∎

Proposition 3.2 *Assume that* $(\mathbf{A}_1)$, ..., $(\mathbf{A}_4)$ *are satisfied. If (3.10) is uniformly stable and attractive at the origin, then it is equi-attractive at the origin.*

Proof. Assume that the origin is not equi-attractive for (3.10). Then, there is an instant t_0 such that for each $\eta > 0$ there exists $\sigma = \sigma(\eta) > 0$ and there exist two sequences $\{x_k\}$, $\{t_k\}$ such that $||x_k|| < \eta$, $t_k > t_0 + k$ and

$$||x(t_k; t_0, x_k)|| \geq \sigma(\eta) \qquad (3.13)$$

for each integer k. Since the origin is attractive for our system, we can associate to such t_0 a number $\delta_0 > 0$ such that

$$\lim_{t \to +\infty} ||x(t; t_0, x)|| = 0 \qquad (3.14)$$

for any $x \in B_{\delta_0}$. Let us choose now $\eta < \delta_0$, and apply Definition 3.2 with $\varepsilon = \sigma(\eta)$. We obtain $\delta = \delta(\eta) > 0$ such that

$$||x|| < \delta(\eta) \implies ||x(t; \tau, x)|| < \sigma(\eta) \qquad (3.15)$$

for each τ and each $t \geq \tau$. Since the sequence $\{x_k\}$ is bounded, it admits a limit point x_0 (without loss of generality, we can assume that $\{x_k\}$ actually converges to x_0). Of course, $||x_0|| < \delta_0$. Hence, by (3.14) there exists $T > 0$ such that for each $t \geq t_0 + T$, we have

$$||x(t; t_0, x_0)|| < \delta(\eta)/2 \; .$$

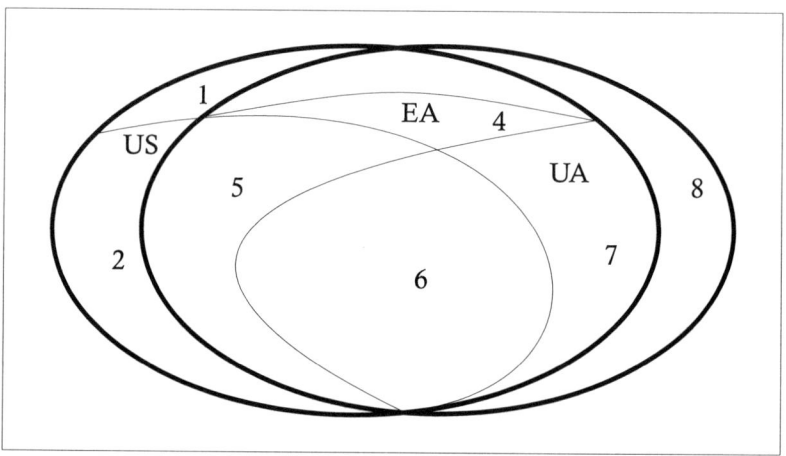

Figure 3.1: Map of stability notions (US=uniform stability, EA=equi-attraction, UA=uniform attraction)

Apply now property (C), and so find $\rho > 0$ such that

$$||x - x_0|| < \rho \Longrightarrow ||x(t_0 + T; t_0, x_0) - x(t_0 + T; t_0, x)|| < \delta(\eta)/2 \ .$$

That is,

$$||x(t_0 + T; t_0, x)|| < \delta(\eta) \ . \tag{3.16}$$

Now, let k be so large that $||x_k - x_0|| < \rho$ and $t_k > t_0 + k > t_0 + T$. Then, (3.16) holds with $x = x_k$ and hence, according to (3.15),

$$||x(t; t_0 + T, x(t_0 + T; t_0, x_k))|| = ||x(t; t_0, x_k)|| < \sigma(\eta)$$

for $t \geq t_0 + T$.

This last inequality must be true in particular for $t = t_k$, but this is impossible because of (3.13). ∎

If we limit ourselves to the class of systems for which the origin is an equilibrium position and $(\mathbf{A}_1)$, ..., $(\mathbf{A}_4)$ are valid, we can therefore represent the relationships among all these notions in the picture of Figure 3.1.

Remark 3.2 None of the regions displayed in Figure 3.1 is nonempty, in general. To show that this is actually the case, we have to point out the existence of at least one explicit example for each region. As far as Regions 2 and 6 are concerned, there are trivial time invariant examples. Time invariant examples exist for Region 8, as well (the most famous one is due to Vinograd, [67] p. 191). In order to construct examples for the other regions, consider the scalar equation

$$\dot{x} = \frac{\dot{f}(t)}{f(t)} x \qquad (3.17)$$

where $f(t)$ is defined and positive for $t \in [0, +\infty)$. Note that any function of the form

$$x(t; t_0, \xi) = \frac{f(t)}{f(t_0)} \xi$$

is a solution of (3.17). If we specify $f(t) = a^{-t \sin^2 t}$ with $a > 1$, we obtain a system which belongs to the Region 1, while taking $f(t) = 1/(t+1)$ we get a system for the Region 5 (see [42] for details). Moreover, if we specify $f(t) = a^{-t(\sin t + 2)}$ ($a > 1$), then the system belongs to the Region 4. Finally, an example for the Region 7 is obtained again in the form (3.17) with $f(t) = a^{6 \sin t - 6t \cos t - t^2}$, with again $a > 1$ (see [98]). Figures 3.2, 3.3, 3.4 present respectively the graphs of the functions $f(t) = a^{-t \sin^2 t}$ (with $a = e$), $f(t) = a^{-t(\sin t + 2)}$ (with $a = e^{1/10}$), $f(t) = a^{6 \sin t - 6t \cos t - t^2}$ (with $a = e^{1/3}$): the numerical values of the bases have been chosen in order to enhance the features of such functions.

The last example concerning the Region 3 is more involved (see [97]). The system is two-dimensional and it is defined in polar coordinates by the equations

$$\dot{\rho} = \frac{\rho}{g(t,\theta)} \frac{\partial g(t,\theta)}{\partial t}, \qquad \dot{\theta} = 0$$

where

$$g(t,\theta) = \frac{\sin^4 \theta}{\sin^4 \theta + (1 - t \sin^2 \theta)^2} + \frac{1}{1 + \sin^4 \theta} \frac{1}{1 + t^2}.$$

∎

The definitions above are classical and all of them are of interest in stability theory. They have been recalled here for sake of completeness, but in this book we focus in particular on uniform stability and uniform asymptotic stability. We need also an updated version of the definition of Lagrange stability (a property often referred to also as uniform boundedness of solutions).

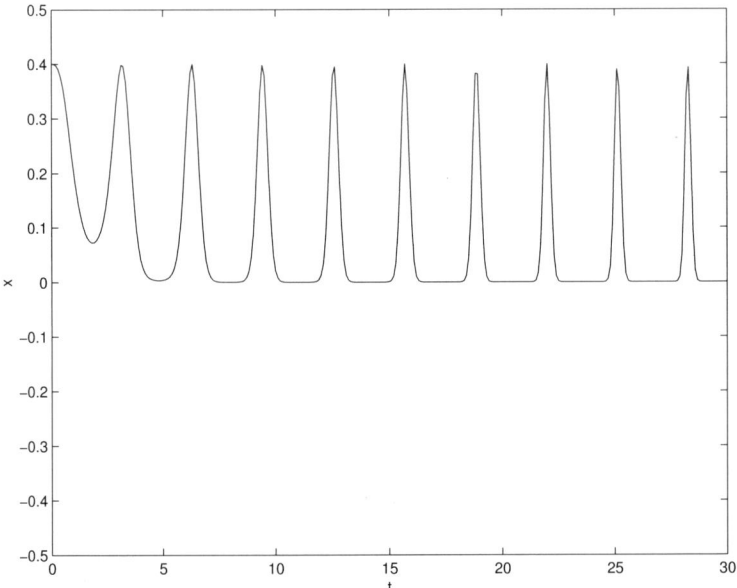

Figure 3.2: Example for region 1

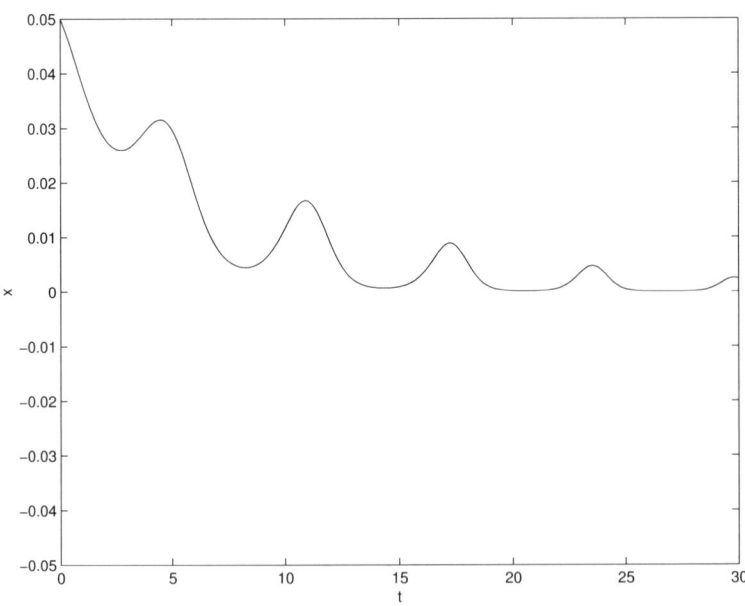

Figure 3.3: Example for region 4

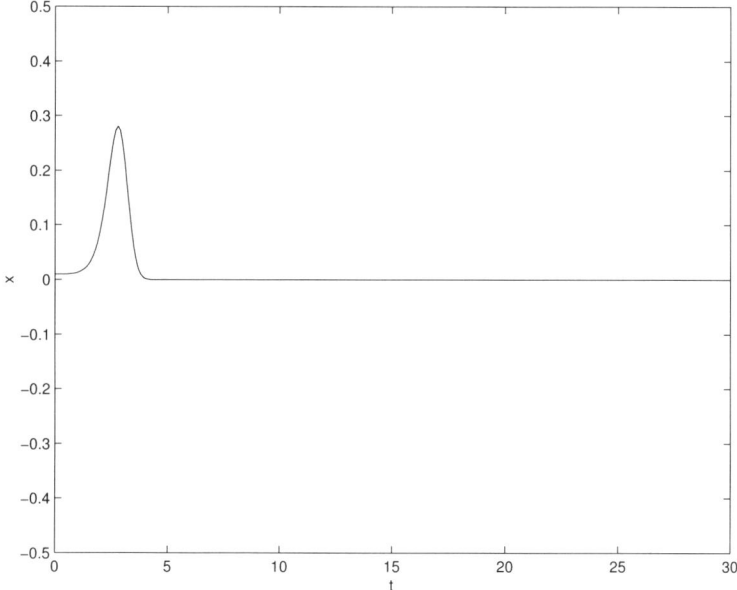

Figure 3.4: Example for region 7

Definition 3.7 *System (3.10) is said to be* uniformly Lagrange stable *if for each $R > 0$ there exists $S > 0$ such that for each pair (t_0, x_0) with $t_0 \geq 0$ and each solution $x(\cdot) \in \mathcal{S}_{t_0, x_0}$ one has that $x(\cdot)$ is continuable on $[t_0, +\infty)$ and that*

$$||x_0|| < R \implies ||x(t)|| < S \quad \text{for each } t \geq t_0 \ .$$

We finally point out that as in the time invariant case, these definitions can be characterized by functions of class $\mathcal{K}$ (see [67], p. 170).

3.2.2 Time dependent Liapunov functions

We introduce now the appropriate generalizations of the definition of Liapunov function (for systems without inputs). We recall that in this section, solutions are intended in Carathéodory sense.

Definition 3.8 *A* weak Liapunov function in the small *for (3.10) is a real map $V(t, x)$ which is defined on $[0, +\infty) \times B_r$ for some $r > 0$, and fulfills the following properties:*

(i) there exist $a, b \in \mathcal{K}_0$ such that

$$a(||x||) \leq V(t,x) \leq b(||x||) \qquad \text{for } t \in [0, +\infty), \; x \in B_r$$

(ii) for each Carathéodory solution $x(\cdot)$ of (3.10) and each interval $I \subseteq [0, +\infty)$ one has

$$t_1, t_2 \in I, \; t_1 < t_2 \Longrightarrow V(t_1, x(t_1)) \geq V(t_2, x(t_2))$$

provided that $x(\cdot)$ is defined on I and $x(t) \in B_r$ for $t \in I$.

It follows from (i) that $V(t,0) = 0$. If there exists a weak Liapunov function in the small for (3.10), then for each t_0, $x(t) \equiv 0$ is the unique solution such that $x(t_0) = 0$. Hence the origin is an equilibrium position.

Definition 3.9 *A function $V(t,x)$ which fulfills the same properties as in Definition 3.8, but with $a, b \in \mathcal{K}^\infty$ instead of $\mathcal{K}_0$, and with B_r replaced by B^r, will be called a* weak Liapunov function in the large *for system (3.10).*

Definition 3.10 *A* strict Liapunov function in the small *for (3.10) is a real map $V(t,x)$ defined on $[0, +\infty) \times B_r$ for some r, which fulfills property (i) and, in addition,*

(iii) there exists $c \in \mathcal{K}_0$ such that for each Carathéodory solution $x(\cdot)$ of (3.10) and each interval $I \subseteq [0, +\infty)$ one has

$$t_1, t_2 \in I, \; t_1 < t_2 \Longrightarrow V(t_2, x(t_2)) - V(t_1, x(t_1)) \leq - \int_{t_1}^{t_2} c(||x(t)||) \, dt$$

provided that $x(\cdot)$ is defined on I and $x(t) \in B_r$ for $t \in I$.

Condition (iii) of Definition 3.10 trivially implies (ii) of Definition 3.8.

Definition 3.11 *A function $V(t,x)$ which fulfills the same properties as in Definition 3.10, but with B_r replaced by $\mathbb{R}^n$, $a, b \in \mathcal{K}_0^\infty$ instead of $\mathcal{K}_0$ and $c \in \overline{\mathcal{K}}_0$, will be called a* global strict Liapunov function *for (3.10).*

Note that if V is differentiable and $f(t,x)$ is continuous, then conditions (ii) and (iii) above can be checked by looking at the sign of the function

$$\dot{V}(t,x) = \frac{\partial V}{\partial t}(t,x) + \sum_{i=1}^{n} \frac{\partial V}{\partial x_i}(t,x) f_i(t,x)$$

(which reduces to $\nabla V(x) \cdot f(x)$ in the time invariant case). Indeed, it is easily seen that

Reformulation of the basic definitions 103

$$(ii) \text{ is equivalent to } \dot{V}(t,x) \le 0$$

and, respectively,

$$(iii) \text{ is equivalent to require } \dot{V}(t,x) \le -c(\|x\|)$$

for each $t \ge 0$ and each $x \in B_r$. However, we emphasize that in the definitions above it is not even required that $V(t,x)$ is continuous.

The function $\dot{V}(t,x)$, when it exists, will be denoted again by the term *derivative of $V(t,x)$ with respect to system* (3.10).

The following comments concern the inequalities in (i). When

$$V(t,x) \ge a(\|x\|) \qquad (3.18)$$

for some $a \in \mathcal{K}_0$ [respectively, $a \in \mathcal{K}^\infty$], we usually say that $V(t,x)$ is *positive definite* [respectively, *radially unbounded*]. This is consistent with the terminology already used in Chapter 2. Indeed, if V is continuous and it does not depend on t (i.e., $V(t,x) = V(x)$), (3.18) with $a \in \mathcal{K}_0$ is equivalent to $V(x) > 0$ for $x \ne 0$ in a neighborhood of the origin, while (3.18) with $a \in \mathcal{K}^\infty$ is equivalent to $\lim_{\|x\| \to +\infty} V(x) = +\infty$ (actually, (3.18) reduces to (2.25)).

The condition $V(t,x) \le b(\|x\|)$ for $b \in \mathcal{K}_0$ is sometimes referred to by saying that V admits an *infinitesimal upper bound*, or that it is *decrescent*. As already noticed, it implies that $V(t,0) = 0$ for each $t \in [0,+\infty)$ and that $V(t,x)$ is continuous at the points of the form $(t,0)$.

Remark 3.3 Inequality (3.18) states that $V(t,x)$ is positive for $x \ne 0$, uniformly with respect to time. It must be emphasized that the weaker requirement

$$V(t,0) = 0 \quad \text{and} \quad V(t,x) > 0 \ (x \ne 0)$$

is not sufficient in general, not even if $V(t,x)$ is continuous. This is shown by the following example. Consider the one-dimensional equation

$$\dot{x} = x \qquad (3.19)$$

which is obviously unstable at the origin. Now take the function $V(t,x)$ defined by

$$V(t,x) = \begin{cases} xe^t & \text{for } x > 0 \\ 0 & \text{for } x = 0 \\ V(t,-x) & \text{for } x < 0 \end{cases}.$$

As required, $V(t,x)$ is positive if $x \neq 0$. Moreover, it is differentiable (at least for $x \neq 0$) and a simple computation shows that

$$\dot{V}(t,x) = e^t x^{e^t}(1 + \log x) .$$

This last expression is negative for $x \in (0, 1/e)$, and hence $V(t,x)$ is decreasing along nontrivial solutions of (3.19). This apparently paradoxical conclusion is explained by the fact that $V(t,x)$ does not admit a uniform estimate from below. ∎

Finally, we point out that not even the uniform estimate from above $V(t,x) \leq b(\|x\|)$ can be suppressed in Definition 3.8 (see the example in [121], p. 27).

3.3 Sufficient conditions

We are now in a position to state the classical first and second Liapunov theorems for the time dependent case, which generalize the analogous theorems presented in Chapter 2. Throughout this section, we still assume that the right hand side of (3.10) meets the conditions $(\mathbf{A}_1)$, $(\mathbf{A}_2)$ and $(\mathbf{A}_3)$.

Theorem 3.1 *Assume that there exists a weak Liapunov function in the small for system (3.10). Then the origin is uniformly stable.*

Theorem 3.2 *Assume that there exists a strict Liapunov function in the small for system (3.10). Then the origin is uniformly locally asymptotically stable.*

If in addition there is a global strict Liapunov function, then the origin is uniformly globally asymptotically stable.

We do not give here the proof of these theorems, which are well known and can be found in many textbooks, for instance in [67]. On the other hand, we notice that Theorem 3.1 is a corollary of next Theorem 3.6 and Proposition 3.3.

As far as Lagrange stability is concerned, we have the following theorem, due to Yoshizawa.

Theorem 3.3 *Assume that there exists a weak Liapunov function in the large for system (3.10). Then the system is Lagrange stable.*

We finally notice that, in spite of the enormous literature about Liapunov method appeared in more than one century, there are still new interesting developments (see for instance [1]).

3.4 Converse theorems

This section is a short survey about converse theorems for stability and uniform stability in the time varying case. As already observed in the time invariant case (Chapter 2), the regularity of $f(t,x)$ does not play a significant role in the proof of direct theorems, but when we deal with converse theorems, it becomes a critical issue. As a consequence, assumptions ($\mathbf{A}_1$), ($\mathbf{A}_2$) and ($\mathbf{A}_3$) about $f(t,x)$ turn out to be no more sufficient in general.

3.4.1 Asymptotic stability

One reason why it is convenient to deal with uniform properties is that (non-uniform) asymptotic stability in general does not imply the existence of any reasonable Liapunov function (see Example 2 in [97], p. 711), not even for systems with analytic right hand side.

The general form of Kurzweil's converse Theorem actually refers to time varying systems and time dependent Liapunov functions (see [91]). The original statement follows: it includes as a particular case Theorem 2.4 of Chapter 2.

Theorem 3.4 *Assume that the origin is an equilibrium position and that the right hand side of (3.10) is continuous with respect to the pair $(t,x) \in [0,+\infty) \times \mathbb{R}^n$. Assume further that the origin is uniformly locally asymptotically stable. Then, there exists a strict Liapunov function in the small of class C^∞.*

In addition, if the origin is uniformly globally asymptotically stable, then it can be found a global strict Liapunov function of class C^∞.

Moreover, if $f(t,x)$ is periodic with respect to t, then the Liapunov function turns out to be periodic with respect to t, and if the system is time invariant then the Liapunov function is time invariant.

The proof of Kurzweil's Theorem will be obtained later as a by-product of a more general result. Note that since in Theorem 3.4 the right hand side of the differential system is assumed to be continuous, then all its solutions are classical.

3.4.2 Uniform stability

As in the time invariant case, when dealing with stability or Lagrange stability the existence of Liapunov functions is more delicate, the critical issue being again the relationship between the regularity of f and the regularity of V.

Earliest studies about converse theorems come back to Persidskii (1936). He actually proved that if a system is stable and if $f(t,x)$ is at least of class C^1, then there exists a C^1 function $V(t,x)$ such that (3.18) holds with $a \in \mathcal{K}_0$ and $\dot{V}(t,x) \leq 0$ for each pair (t,x) (which of course implies property (ii) of Definition 3.8). Moreover, Persidskii introduced the notion of uniform stability and gave the first proof of Theorem 3.1. The converse of Theorem 3.1 was obtained later, around 1955, independently by Krasovski ([88]), Kurzweil ([90]) and Yoshizawa ([158]). It should be emphasized that Kurzweil's construction applies when f is of class C^1, while for Yoshizawa's construction, f continuous and locally Lipschitz continuous with respect to x, suffices. The precise statement of Yoshizawa's result is as follows.

Theorem 3.5 *Consider the system (3.10), and assume that $f(t,x)$ is continuous on $[0, +\infty) \times \mathbb{R}^n$. Assume further that it is locally Lipschitz continuous with respect to x. If the origin is uniformly stable, then, there exists a weak Liapunov function in the small which is locally Lipschitz continuous with respect to both t and x.*

If in addition $f(t,x)$ is locally Lipschitz continuous with respect to both t and x, then there exists a weak Liapunov function in the small of class C^∞.

Remark 3.4 Having in mind the last sentence in the statement of Theorem 3.4, one can be tempted to conjecture that if f is sufficiently regular, periodic with respect to t (or independent of t) and uniformly stable, then it is possible to find a periodic smooth Liapunov function. This is false, as the following argument shows.

Consider a time invariant stable system

$$\dot{x} = f(x) \tag{3.20}$$

which does not admit a continuous Liapunov function (e.g., the system in Example 2.2), and assume that we can find a continuous, weak Liapunov function $V(t,x)$, periodic with respect to t. Let $T > 0$ be a period for $V(t,x)$. Let us define

$$W(x) = \int_0^T V(s,x)\,ds \ . \tag{3.21}$$

It is clear that $W(x)$ is positive definite and continuous. Let us prove that for any solution $x(t)$ such that $x(0) = \bar{x}$ and for each $t \geq 0$, we have

$$W(x(t)) \leq W(\bar{x}) \ .$$

To this end, let $x_s(t) = x(t-s)$. Of course, $x_s(t)$ is a solution of (3.20) such that $x_s(s) = \bar{x}$. Therefore, for each $t \geq 0$ and each $s \in [0,T]$ we have

$$V(s,\bar{x}) = V(s, x_s(s)) \geq V(s+t, x_s(s+t)) = V(s+t, x(t)) \ .$$

Hence,

$$\begin{aligned} W(\bar{x}) &\geq \int_0^T V(s+t, x(t))\,ds = \int_t^{t+T} V(\sigma, x(t))\,d\sigma \\ &= \int_0^T V(\sigma, x(t))\,d\sigma = W(x(t)) \ . \end{aligned}$$

In conclusion, $W(x)$ is a continuous, weak Liapunov function for our system, which is a contradiction.

Note that (3.21) allows us to define a time invariant Liapunov function $W(x)$ which is as regular as $V(t,x)$ is (continuous, Lipschitz continuous or C^r with $1 \leq r \leq +\infty$). ∎

As a consequence of Theorem 3.5, we see that a continuous time dependent Liapunov function must exist for the center-focus configuration considered in Chapter 2 (Example 2.2) and for its one-dimensional version. In fact, a continuous Liapunov function has been explicitly found in Example 3.1. The construction performed in Example 3.1 illustrates Yoshizawa's idea on which the proof of Theorem 3.5 is based.

If the system is stable and f is merely continuous, the same construction can be adapted and gives rise again to a weak Liapunov function in the small: however, in general, there exists no continuous such a Liapunov function. This is shown by a counter-example due to Kurzweil and Vrkoč (see [92]). We present here a slightly modified version.

Example 3.3 Consider the time invariant one-dimensional equation

$$\dot{x} = f(x) \tag{3.22}$$

where

$$f(x) = \begin{cases} 2^n \cdot \sqrt{1 - (2^{-n}x - 3)^2} & \text{for } 3 \cdot 2^n < x \leq 2^{n+2} \\ 2^n \cdot \sqrt{(1 - (2^{-n}x - 3)^2)^3} & \text{for } 2^{n+1} < x \leq 3 \cdot 2^n \end{cases}, \quad n \in \mathbb{Z}$$

$f(0) = 0$ and $f(-x) = -f(x)$. Note that the right hand side of (3.22) is continuous, but not Lipschitz continuous.

The behavior of solutions can be described in the following way. For $x > 0$, there are infinitely many equilibrium positions P_n, $n \in \mathbb{Z}$. The sequence diverges when $n \to +\infty$, and converges to the origin when $n \to -\infty$. The corresponding graphs divide the region $x > 0$, $t > 0$ into infinitely many strips. Any solution whose graph is contained in one of these strips is increasing and collapses with the graph of the upper equilibrium solution in finite time.

This situation looks like the one-dimensional version of center-focus configuration, but with the fundamental difference that in the latter case the solutions do not collapse: they approach the upper one in an asymptotic way.

It is evident that this system is uniformly stable at the origin and Lagrange stable. It should be also clear that a continuous (time varying) Liapunov function cannot exist for this system.

3.5 Robust stability

Kurzweil and Vrkoč proved in [92] that the existence of a continuous weak Liapunov function becomes necessary and sufficient when the definition of stability is conveniently strengthened. We report here their definition, which will be referred to as "robust stability"[1]. Also in this section, solutions are always intended in Carathéodory sense, and conditions ($\mathbf{A_1}$), ($\mathbf{A_2}$) and ($\mathbf{A_3}$) are assumed to hold.

Definition 3.12 *We say that (3.10) is robustly locally stable at the origin if there exist a sequence $\{G_i\}_{i=0,1,2,...}$ of open sets in $[0,+\infty) \times \mathbb{R}^n$ and two sequences of real numbers $\{a_i\}_{i=0,1,2,...}$ and $\{b_i\}_{i=0,1,2,...}$ such that*

(I) $0 < b_{i+1} < a_i \leq b_i$ for each $i = 0, 1, 2, \ldots$, and $b_i \to 0$ for $i \to +\infty$

(II) $[0,+\infty) \times \{x : |x| < a_i\} \subseteq G_i \subseteq [0,+\infty) \times \{x : |x| < b_i\}$ for each $i = 0, 1, 2, \ldots$

(III) for each $i = 0, 1, 2, \ldots$, each initial pair $(t_0, x_0) \in G_i$ and each solution $x(\cdot) \in \mathcal{S}_{t_0,x_0}$ one has that $x(\cdot)$ is continuable on $[t_0,+\infty)$ and $(t, x(t)) \in G_i$ for each $t \geq t_0$.

---

[1] The term "robust stability" is often used in control theory to denote stability with respect to external disturbances or uncertainty of structural parameters, as in previous Section 2.2.5. From now on, robust stability will be rather used with the meaning of Definition 3.12. Thus, no confusion is possible.

Next two propositions clarify the relations between uniform stability and robust stability.

Proposition 3.3 *If the origin is robustly stable for (3.10), then it is uniformly stable.*

Proof. Let $\varepsilon > 0$ be given, and let $i \in \mathbb{N}$ be such that $b_i < \varepsilon$. We claim that the choice $\delta = a_i$ works. Indeed, for any $t_0 \geq 0$ and any x_0 with $||x_0|| < \delta$, we have

$$(t_0, x_0) \in [0, +\infty) \times \{x : ||x|| < a_i\} \subseteq G_i .$$

According to (III), this implies that $(t, x(t)) \in G_i$ for each $x(\cdot) \in \mathcal{S}_{t_0, x_0}$ and each $t \geq t_0$. Hence, $||x(t)|| < b_i < \varepsilon$ for each $t \geq t_0$. ∎

Example 3.3 shows that the converse of Proposition 3.3 is false, in general.

Proposition 3.4 *Assume that $f(t, x)$ satisfies condition ($\mathbf{A}_4$) (local Lipschitz continuity with respect to x), besides ($\mathbf{A}_1$), ($\mathbf{A}_2$) and ($\mathbf{A}_3$). Then, uniform stability at the origin implies robust stability.*

Proof. We start by setting $b_0 = 1$. According to Definition 3.2, to $\varepsilon = b_0$ there corresponds a number $\delta > 0$. We take $a_0 = \delta$.

It is clear that $a_0 \leq b_0$. Next, we define G_0 as the set of all points of the form $(t, x(t))$ with $x(\cdot) \in \mathcal{S}_{t_0, x_0}$, $t \geq t_0 \geq 0$ and $||x_0|| < a_0$.

It follows immediately that

$$[0, +\infty) \times \{x : ||x|| < a_0\} \subset G_0 \subset [0, +\infty) \times \{x : ||x|| < b_0\} .$$

It remains to prove that G_0 is open. To this purpose, we can use property (C) (which is guaranteed under our assumptions) and the fact that $[0, +\infty) \times \{x : ||x|| < a_0\}$ is open.

Assume now that a_i, b_i and G_i have been constructed. We can choose any $b_{i+1} \in (0, a_i)$ and repeat inductively the previous construction. ∎

We state now the analogous of first Liapunov's theorem for robust stability.

Theorem 3.6 *Assume that there exists a continuous weak Liapunov function in the small for (3.10). Then, (3.10) is robustly stable at the origin.*

Proof. First of all we observe that under the assumptions of the theorem any solution issuing from a sufficiently small neighborhood of the origin is right continuable to $+\infty$. To prove it, let r_0 be any positive number such that $a(r_0)$ is defined. Since $b(r) \to 0$ for $r \to 0^+$, there exists $\delta_0 > 0$ such that

$$a(r_0) \geq b(\delta_0) .$$

Without loss of generality, we can assume that $\delta_0 < r_0$. Let $\|x_0\| < \delta_0$. We claim that for each $t_0 \geq 0$, each $x(\cdot) \in \mathcal{S}_{t_0,x_0}$ and each $t \geq t_0$ one has $x(t) \in B_{r_0}$. In the opposite case, we should have

$$\|x(t_1)\| \geq r_0$$

for some $t_1 > t_0$. Hence,

$$V(t_1, x(t_1)) \geq a(\|x(t_1)\|) \geq a(r_0) \geq b(\delta_0) > b(\|x_0\|) \geq V(t_0, x_0)$$

which is a contradiction to Definition 3.8(ii).

Next we define G_i, a_i and b_i for $i = 0$. Let $b_0 = \delta_0$ and let $\lambda_0 = a(b_0)$. Since $b(0) = 0$, and $\lambda_0 = a(\delta_0) \leq b(\delta_0)$, it makes sense to take $a_0 = b^{-1}(\lambda_0)$. We finally set

$$G_0 = \{(t, x) : V(t, x) < \lambda_0\} .$$

Since V is continuous, G_0 is open in $[0, +\infty) \times \mathbb{R}^n$. Clearly, $a_0 \leq b_0$. Moreover, by construction, we have $b_0 = \delta_0 < r_0$.

If $\|x\| < a_0$, then $b(\|x\|) < \lambda_0$. Hence, for each $t \geq 0$, $V(t, x) < \lambda_0$. On the other hand, if $V(t, x) < \lambda_0$ for some pair (t, x), then $a(\|x\|) < \lambda_0$, so that $\|x\| < a^{-1}(\lambda_0) = \delta_0 = b_0$. This shows that G_0 satisfies condition (II). Finally, condition (III) follows from the definition of G_0 and property (ii) of Definition 3.8. In particular, we note that all the solutions starting from G_0 are right continuable to $+\infty$.

Now, assume that

$$a_0, a_1, \ldots, a_{i-1}, \qquad b_0, b_1, \ldots, b_{i-1}, \qquad G_0, G_1, \ldots, G_{i-1}$$

have been defined for some $i \geq 1$, in such a way that

$$0 < a_{i-1} \leq b_{i-1} < \ldots < a_1 \leq b_1 < a_0 \leq b_0 ,$$

G_j is open in $[0, +\infty) \times \mathbb{R}^n$ and (II), (III) hold ($j = 0, 1, \ldots, i - 1$)

and, in addition,
$$a_j \le \frac{r_0}{2^j} \quad (j = 0, 1, \ldots, i-1) \ .$$

Let us define
$$\lambda_i = \min\left\{a\left(\frac{a_{i-1}}{2}\right), b\left(\frac{r_0}{2^i}\right)\right\} , \qquad (3.23)$$
$a_i = b^{-1}(\lambda_i)$, $b_i = a^{-1}(\lambda_i)$ and finally $G_i = \{(t, x) : V(t, x) < \lambda_i\}$. Of course, $a_i \le b_i$. Moreover, according to (3.23), we have
$$b_i = a^{-1}(\lambda_i) \le \frac{a_{i-1}}{2} < a_{i-1} \ .$$
Using again (3.23), we have also $b^{-1}(\lambda_i) \le r_0/2^i$ and hence
$$a_i \le \frac{r_0}{2^i} \ . \qquad (3.24)$$
To prove that the open set G_i meets conditions (II) and (III), we proceed as in the case $i = 0$.

To complete the proof, it remains to notice that a_i and b_i converge to zero, as required, by virtue of (3.24). ∎

Under the assumption that $f(t, x)$ is continuous, Kurzweil and Vrkoč proved in [92] that also the converse of Theorem 3.6 holds.

Theorem 3.7 *Let $f(t, x)$ be continuous on $[0, +\infty) \times \mathbb{R}^n$. Assume that (3.10) is robustly locally stable at the origin. Then, there exists a C^∞ weak Liapunov function in the small.*

It is worth noticing that the Liapunov function provided by Kurzweil and Vrkoč converse theorem is of class C^∞. We do not report the proof of Theorem 3.7. Indeed, even in this case a more general statement will be given and proved in the next chapter, in the more general context of systems defined by differential inclusions.

Remark 3.5 It should be noticed that the new restrictions introduced in the definition of robust stability, when applied to the case of time invariant systems, turn out to be weaker than the conditions about prolongations (absolute stability) discussed in Chapter 2. Consider for instance the system in Example 3.1. A continuous, time dependent Liapunov function was explicitly constructed. Hence, from Theorem 3.6, the system is robustly stable at the origin. However, we know that a continuous time invariant weak Liapunov function cannot exist, so that the system is not absolutely stable at the origin.

3.6 Lagrange stability

In analogy with the case of uniform stability, Yoshizawa proved that a conclusion similar to that of Theorem 3.5 is valid also for uniform Lagrange stability.

Theorem 3.8 *Consider the system (3.10), and assume that $f(t,x)$ is continuous on $[0,+\infty) \times \mathbb{R}^n$, and locally Lipschitz continuous with respect to x (condition $(\mathbf{A}_4)$). Assume further that the system is Lagrange stable. Then, there exists a weak Liapunov function in the large which is locally Lipschitz continuous with respect to both t and x.*

If in addition $f(t,x)$ is locally Lipschitz continuous with respect to the pair (t,x), then there exists a weak Liapunov function in the large of class C^∞.

Example 3.3 can be used to show that Lagrange stability does not imply the existence of a continuous weak Liapunov function in the large, if the right hand side of the system is assumed to be only continuous.

The existence of a continuous weak Liapunov function in the large turns out to be actually equivalent to a stronger type of Lagrange stability, that we agree to call "robust Lagrange stability".

Definition 3.13 *We say that (3.10) is robustly Lagrange stable if there exist a sequence $\{G_i\}_{i=0,1,2,...}$ of open sets in $[0,+\infty) \times \mathbb{R}^n$ and two sequences of real numbers $\{a_i\}_{i=0,1,2,...}$ and $\{b_i\}_{i=0,1,2,...}$ such that*

(I) $0 < a_i \le b_i < a_{i+1}$ for each $i = 0,1,2,\ldots$, and $a_i \to +\infty$ for $i \to +\infty$

(II) $[0,+\infty) \times \{x : ||x|| < a_i\} \subseteq G_i \subseteq [0,+\infty) \times \{x : ||x|| < b_i\}$ for each $i = 0,1,2,\ldots$

(III) for each $i = 0,1,2,\ldots$, for each initial pair $(t_0, x_0) \in G_i$ and each solution $x(\cdot) \in \mathcal{S}_{t_0,x_0}$, one has that $x(\cdot)$ is continuable on $[0,+\infty)$ and $(t,x(t)) \in G_i$ for each $t \ge t_0$.

As in the case of local stability (see Propositions 3.3 and 3.4) it is possible to prove that the robust Lagrange stability implies the usual Lagrange stability property. The two properties are actually equivalent when f satisfies conditions $(\mathbf{A}_1), \ldots, (\mathbf{A}_4)$.

Theorem 3.9 *Assume that there exists a continuous weak Liapunov function in the large for (3.10), Then, (3.10) is robustly Lagrange stable.*

Apart from some obvious modifications, the proof of Theorem 3.9 is very similar to the proof of Theorem 3.6 and it is left to the reader.

3.7 Discontinuous right hand side

In the previous sections we considered systems for which at least the conditions ($\mathbf{A}_1$), ($\mathbf{A}_2$) and ($\mathbf{A}_3$) are satisfied. Solutions were intended in Carathéodory sense and the stability notions were consistently defined.

Next step is to remove assumption ($\mathbf{A}_3$). In other words, we would like to extend, as far as possible, the results illustrated so far to the case where $f(t,x)$ is locally bounded and, say, Lebesgue measurable with respect to both variables. As explained in Chapter 1, systems with discontinuous right hand side are often treated as differential inclusions. Although the Liapunov method for differential inclusions will be fully exploited as the main subject of next chapter, we find it convenient to anticipate here some comments and one result.

It has been recently proved in [116] that Theorem 3.4 (Kurzweil's Theorem) extends to discontinuous systems, provided that solutions are intended in Filippov's sense.

Let us assume that the definitions of stability and asymptotic stability, as well as the definitions of Liapunov function, have been updated by replacing Carathéodory solutions by Filippov solutions. Note that in order to have an equilibrium position at the origin, $x(t) \equiv 0$ must be now a Filippov solution. This is equivalent to require $0 \in \mathbf{K}_x f(t, 0)$ for a.e. $t \geq 0$.

Theorem 3.10 *Assume that the origin is an equilibrium position and that the right hand side of (3.10) is locally bounded and measurable with respect to the pair $(t, x) \in [0, +\infty) \times \mathbb{R}^n$. Assume further that the origin is locally uniformly asymptotically stable. Then, there exists a strict Liapunov function in the small which is locally Lipschitz continuous.*

In addition, if the origin is globally uniformly asymptotically stable, then it can be found a locally Lipschitz continuous global strict Liapunov function.

Finally, if the system is time invariant (i.e., $f(t,x) = f(x)$) then the Liapunov function is time invariant.

For the proof of Theorem 3.10 we refer to [116]. Note that the statement about the regularity of $V(t,x)$ cannot be improved in general. An example will be given in Chapter 4.

3.8 Time varying feedback

It has been shown in the first section of this chapter (Example 3.1) that there are systems which can be stabilized by a continuous, time varying feedback, but not by a continuous, time invariant one. In the present section we review some results by J.M. Coron ([46], [47]). Basically, we shall see that stabilizability under time dependent (actually periodic) feedback is possible for each system, provided that the dimension of the state space is sufficiently high and suitable controllability conditions are satisfied.

First of all, we need a further refinement of the notion of STLC introduced in Section 2.3.2. Here, an admissible control on a given interval $[0,T]$ will be thought of as an element of the normed space $L_1([0,T],\mathbb{R}^n)$. We shall say that a time invariant system

$$\dot{x} = f(x,u) \qquad (3.25)$$

is *continuously* STLC if it is STLC with small controls and, in addition, the map $x \mapsto u_x(t)$ from a neighborhood of the origin of $\mathbb{R}^n$ to $L_1([0,T],\mathbb{R}^n)$ is continuous. It is proven in [43] that many well known computable sufficient conditions for STLC allow us to conclude that the system is actually continuously STLC. The simplest way to state Coron's result is the following one.

Theorem 3.11 *Let $n \geq 4$. Assume that $f(x,u)$ is real analytic on $\mathbb{R}^n \times \mathbb{R}^m$, and assume that (3.25) is continuously STLC. Then, for each $T > 0$ there exist $\varepsilon > 0$ and a function $(t,x) \mapsto u(t,x)$ which enjoys the following properties:*

- *$u(t,x)$ is defined and continuous on $\mathbb{R} \times \mathbb{R}^n$*
- *$u(t,x)$ is of class C^∞ on $\mathbb{R} \times (\mathbb{R}^n \setminus \{0\})$*
- *for each $x \in \mathbb{R}^n$ and each $t \in \mathbb{R}$, $u(t+T,x) = u(t,x)$*
- *for each solution $x(t)$ of the closed loop system*

$$\dot{x} = f(x,u(t,x)) \,, \qquad (3.26)$$

if $x(\tau) = 0$ for some $\tau \in \mathbb{R}$, then $x(t) = 0$ for each $t \geq \tau$

- *for each solution $x(t)$ of (3.26), if $||x(\tau)|| < \varepsilon$ for some $\tau \in \mathbb{R}$, then $x(t) = 0$ for each $t \geq \tau + T$.*

It follows from the properties listed above that (3.26) is locally uniformly asymptotically stable at the origin ([46], Lemma 2.15).

In other words, Coron's result states that for analytic systems and for $n \geq 4$, a special notion of local controllability implies continuous stabilizability (in fact, almost smooth stabilizability).

Note that the periodic feedback $u(t,x)$ in Theorem 3.11 guarantees in some sense more than asymptotic stabilizability: the origin is exactly reached in finite time.

Coron's result remains valid also when $n = 1$. In fact, in this case it is possible to construct a time invariant feedback ([45]). Coron's result is true without any restriction on the state space dimension in the case of a driftless affine systems ([44]). However, in the general case it is not known whether Coron's result is valid for $n = 2, 3$.

The statement of Theorem 3.11 can be extended to the case where f is of class C^∞, provided that an extra assumption (implicitly fulfilled in the analytic case) is explicitly made. This assumption has a Lie algebraic nature. In order to describe it, we need to introduce some notation. By $\alpha \in \mathbb{N}^m$, we mean a multi-index $\alpha = (\alpha_1, \ldots, \alpha_m)$ and by $|\alpha|$ we mean its length, $|\alpha| = \alpha_1 + \ldots + \alpha_m$. Moreover, if $\mathcal{V}$ and $\mathcal{W}$ are sets of vector fields, by $[\mathcal{V}, \mathcal{W}]$ we mean the set of all vector fields of the form $[X, Y]$, with $X \in \mathcal{V}$, $Y \in \mathcal{W}$ (as usual, $[\cdot, \cdot]$ denotes the Lie bracket).

We consider some families of vector fields associated to the system (3.25):

$$\mathcal{D}_0 = \left\{ \frac{\partial^{|\alpha|}}{\partial u^\alpha} f(\cdot, 0) \, , \, \alpha \in \mathbb{N}^m \right\}$$

$$\mathcal{D}_1 = \left\{ \frac{\partial^{|\alpha|}}{\partial u^\alpha} f(\cdot, 0) \, , \, \alpha \in \mathbb{N}^m \, , \, |\alpha| \geq 1 \right\}$$

$$\mathcal{E}_2 = \{[\mathcal{D}_0, \mathcal{D}_0]\}$$

$$\mathcal{E}_3 = \{[\mathcal{D}_0, \mathcal{E}_2]\}$$

....

$$\mathcal{E} = \cup_{i=2}^\infty \mathcal{E}_i$$

and, finally,

$$\mathcal{D}(0) = \text{span}\{X(0) : X \in \mathcal{D}_1 \cup \mathcal{E}\} \, .$$

Note that $\mathcal{E}$ represents the set of all iterated Lie brackets of elements of $\mathcal{D}_0$, but in general $\mathcal{D}_0$ is not contained in $\mathcal{E}$.

We are now ready to state Coron's Theorem in its generality.

Theorem 3.12 *Let $n \geq 4$. Assume that $f(x, u)$ is of class C^∞ on $\mathbb{R}^n \times \mathbb{R}^m$, and assume that (3.25) is continuously STLC. Assume finally that $\dim \mathcal{D}(0) = n$. Then the conclusions of Theorem 3.11 remain valid.*

If f is analytic, or a polynomial with respect to u, then the continuous STLC assumption implies that $\dim \mathcal{D}(0) = n$. The more common criteria for STLC actually imply that $\dim \mathcal{D}(0) = n$.

The first pioneering work about time dependent periodic stabilizing feedback is due to Sontag and Sussmann ([139]). We also report the following theorem ([49]), relating time varying stabilization and discontinuous stabilization.

Theorem 3.13 *Let $f : \mathbb{R}^n \times \mathbb{R}^m \to \mathbb{R}^n$ be continuous, and let $f(0,0) = 0$. Assume that (2.27) can be stabilized in Filippov's sense by means of a discontinuous feedback law fulfilling (2.44). Then, for each positive number T, (2.27) can also be stabilized by means of a continuous time varying (periodic) feedback law of period T.*

We conclude this section by a remark concerning Artstein's example discussed in Chapter 2. We prove that it cannot be stabilized by a continuous time varying feedback.

Proposition 3.5 *The Artstein's system*

$$\begin{cases} \dot{x}_1 = (x_1^2 - x_2^2)u, \\ \dot{x}_2 = 2x_1 x_2 u, \end{cases} \quad (3.27)$$

is not stabilizable by means of a continuous time varying feedback law $u = u(t,x)$.

Proof[2]. Arguing by contradiction, we assume that there exists a continuous time varying feedback $u(t,x)$, which (locally) stabilizes (3.27) at the origin. Thanks to Theorem 3.4 we see that u may be smooth out for $x \neq 0$. Therefore we may assume that the vector field $f(t,x) := \left((x_1^2 - x_2^2)u(t,x), 2x_1 x_2 u(t,x)\right)^T$ is in $C^\infty(\mathbb{R}^+ \times (\mathbb{R}^n \setminus \{0\}))$, which, together with the stability of the origin, implies that the solutions of

$$\dot{x} = f(t,x) \quad (3.28)$$

are *unique* in forward time. It follows that the corresponding flow map $\varphi(t, t_0, x_0)$ is jointly continuous. (Here, $\varphi(\cdot, t_0, x_0)$ denotes the solution of (3.28) issuing from x_0 at $t = t_0$.)

Set, for any $t \in [0,1]$ and any $x \in B_r$ ($r > 0$ small enough),

$$H(t,x) := \begin{cases} \varphi\left(\frac{t}{1-t}, 0, x\right) & \text{if } 0 \leq t < 1, \\ 0 & \text{if } t = 1. \end{cases}$$

---

[2]Communicated to the authors by J.M. Coron.

Since the origin is (uniformly) asymptotically stable for (3.28), it is clear that H is also jointly continuous. Pick any number $\rho < r/2$ and set $\bar{x} := (0, \rho)^T$, $C := \partial B_\rho(\bar{x}) \,(\subset B_r)$. The circle C is left invariant by the flow of (3.28), since $f(t,x)$ is tangent to C for any $(t,x) \in \mathbb{R}^+ \times C$. Thus H maps continuously $[0,1] \times C$ into C and it fulfills $H(0,x) = x$, $H(1,x) = 0$ for all $x \in C$. It means that the circle C is contractile, which is absurd. ∎

Chapter 4

Differential inclusions

In this chapter, we are concerned with stability properties for a differential inclusion $\dot{x} \in F(t,x)$. We first investigate the (uniform) global asymptotic stability. After a brief review of the main contributions in this topic, we give a complete (and short) proof of a result which generalizes and unifies all the converses of the second Liapunov theorem in the literature. The second part of the chapter is devoted to the first Liapunov theorem and its converse (given together with its proof) for differential inclusions.

4.1 Global asymptotic stability

The reader is referred to Chapter 1 for the notations (namely, $\|\cdot\|$, $\overline{B}_r(x)$, $h(A,B)$, etc.) and the current assumptions (($\mathbf{H}_1$),...,($\mathbf{H}_4$), Hausdorff continuity, etc.), and to Chapter 3 for some definitions of stability, whose extension to differential inclusions is straightforward.

4.1.1 Sufficient conditions

Let $F: \mathbb{R}^+ \times \mathbb{R}^n \to \mathbb{R}^n$ be a multivalued map, fulfilling ($\mathbf{H}_1$), ($\mathbf{H}_2$), ($\mathbf{H}_3$) and ($\mathbf{H}_4$). It follows that for each pair $(t_0, x_0) \in \mathbb{R}^+ \times \mathbb{R}^n$, a solution $x(\cdot)$ of the differential inclusion

$$\dot{x} \in F(t,x), \qquad (4.1)$$

issuing from x_0 at $t = t_0$, exists, at least *locally*. (The set of the maximal solutions of above Cauchy problem is still denoted by $\mathcal{S}_{t_0, x_0}$.) Assume that the origin is an equilibrium position for (4.1), i.e., $0 \in F(t,0)$ for a.e. $t \geq 0$. We aim to characterize the (possible) asymptotic stability of the origin by means of some (strict) Liapunov function. According to Definition 3.5, (4.1) is said to be (uniformly) globally asymptotically stable (UGAS, in short), if the origin is uniformly stable (and Lagrange stable) and uniformly attractive. In other words, it means that there exist two functions $m\colon (0,+\infty) \to (0,+\infty)$ and $T\colon (0,+\infty)^2 \to (0,+\infty)$, such that

(i) for each $R > 0$, each pair $(t_0, x_0) \in \mathbb{R}^+ \times \mathbb{R}^n$ and each $x(\cdot) \in \mathcal{S}_{t_0, x_0}$

$$\|x_0\| \leq R \quad \Rightarrow \quad \|x(t)\| < m(R) \quad \forall t \geq t_0;$$

(ii) $\lim_{R \to 0^+} m(R) = 0$;

(iii) for each $R > 0$, each $\varepsilon > 0$, each pair $(t_0, x_0) \in \mathbb{R}^+ \times \mathbb{R}^n$ and each $x(\cdot) \in \mathcal{S}_{t_0, x_0}$

$$\|x_0\| \leq R \quad \Rightarrow \quad \|x(t)\| < \varepsilon \quad \forall t \geq t_0 + T(R, \varepsilon).$$

The following result is an obvious generalization of Proposition 2.11. Its proof may be found in [91, Lem. 1].

Proposition 4.1 *The following statements are equivalent:*
(i) The origin is UGAS for (4.1);
(ii) there exists a function $\beta \colon [0, +\infty) \times [0, +\infty) \to \mathbb{R}^+$ which is of class $\mathcal{LK}$ and such that

$$\|x(t_0 + h)\| \leq \beta(h, \|x_0\|)$$

for each $t_0 \in \mathbb{R}^+$, each $x_0 \in \mathbb{R}^n$, each $x(\cdot) \in \mathcal{S}_{t_0, x_0}$ and each $h \geq 0$.

The proof of the following theorem is a straightforward extension of the proof of the classical Theorem 3.2.

Theorem 4.1 (SECOND LIAPUNOV THEOREM) *Let $F \colon [0, +\infty) \times \mathbb{R}^n \to \mathbb{R}^n$ be a set-valued map such that the (local) existence of solutions of (4.1) is insured. Assume that there exists a strict Liapunov function in the large V, i.e., a function $V = V(t, x)$ such that, for some functions $a, b, c \in \mathcal{K}_0^\infty$,*

$$a(\|x\|) \leq V(t, x) \leq b(\|x\|) \quad \text{for all } t \in [0, +\infty),\ x \in \mathbb{R}^n, \qquad (4.2)$$

$$t_1 \leq t_2 \ \Rightarrow\ V(t_2, x(t_2)) - V(t_1, x(t_1)) \leq -\int_{t_1}^{t_2} c(\|x(\tau)\|)\, d\tau \qquad (4.3)$$

for each pair of times (t_1, t_2) and each solution $x(\cdot) \colon [t_1, t_2] \to \mathbb{R}^n$ of (4.1). Then the origin is UGAS for (4.1).

When V is of class C^1, for condition (4.3) to be fulfilled it is sufficient (not necessary[1]) that the following (strong) infinitesimal decrease condition holds:

$$\frac{\partial V}{\partial t}(t,x) + \langle \nabla_x V(t,x), v \rangle \leq -c(\|x\|)$$

for a.e. $t \geq 0$, for all $x \in \mathbb{R}^n$ and all $v \in F(t,x)$.

(Here, $\langle \cdot, \cdot \rangle$ denotes the usual scalar product in $\mathbb{R}^n$, and $\nabla_x V := \left(\frac{\partial V}{\partial x_1}, ..., \frac{\partial V}{\partial x_n} \right)^T$.)

Let us end this section by some useful result.

Proposition 4.2 *Let $F : \mathbb{R}^+ \times \mathbb{R}^n \to \mathbb{R}^n$ be a locally Lipschitz continuous (for the Hausdorff distance) multivalued map, which assumes nonempty closed values for each (t,x). If (4.1) is UGAS at the origin, then the same is true for the system $\dot{x} \in \overline{\mathrm{co}}\{F(t,x)\}$.*

Proof. It is sufficient to go back to the definition of the (uniform) global asymptotic stability for (4.1), expressed in terms of $m(R)$ and $T(R,\varepsilon)$ (as above), and to invoke [59, Thm. 3]. ∎

4.1.2 Time invariant systems

In this subsection, we limit ourselves to the time independent case, i.e., we consider differential inclusions of the form

$$\dot{x} \in F(x) , \qquad (4.4)$$

where $F : \mathbb{R}^n \to \mathbb{R}^n$ takes nonempty compact values. The first converse of second Liapunov theorem in this context has been given by Lin, Sontag and Wang in [95, Thm 2.9].

Theorem 4.2 *Assume that the origin is UGAS for (4.4), where F is a locally Lipschitz continuous multivalued map, which takes nonempty compact values. Then there exists a C^∞ global strict Liapunov function V, which satisfies*

$$\langle \nabla_x V(x), v \rangle \leq -c(\|x\|) \qquad \forall x \in \mathbb{R}^n, \; \forall v \in F(x),$$

for some function $c \in \mathcal{K}_0^\infty$.

Actually, Theorem 4.2 is stated in [95] in a somewhat different manner: (i) F takes in [95] the special form $F(x) := \{f(x,d), d \in D\}$, where $f : \mathbb{R}^n \times \mathbb{R}^m \to \mathbb{R}^n$

---
[1] If F is not Lipschitz continuous, there may exist a vector $v \in F(t,x)$ which cannot be written as a derivative $\dot{x}(t)$ for some solution x of (4.1).

is a smooth function and D is a compact set in $\mathbb{R}^m$; (ii) Theorem 2.9 in [95] deals with the asymptotic stability with respect to any *compact invariant set*, instead of the origin. Theorem 4.2 has been used in [142] to prove that the converse of Theorem 2.15 holds true, as well.

Another converse Liapunov theorem has been obtained a few years later by Clarke, Ledyaev and Stern in [38] for another class of multivalued maps.

Theorem 4.3 *Assume that the origin is UGAS for (4.4), where F is an upper semi-continuous (u.s.c.) multivalued map, which takes nonempty compact convex values. Then there exists a C^∞ global strict Liapunov function V, which satisfies $\langle \nabla_x V(x), v \rangle \leq -W(x)$ for each $x \in \mathbb{R}^n$ and each $v \in F(x)$, for some definite positive continuous function W.*

Here, F is only supposed to be u.s.c., but $F(x)$ is convex for each x. Thanks to Proposition 4.2, Theorem 4.2 turns out to be a direct consequence of Theorem 4.3.

Finally, we recall the recent paper [148], where the authors are interested in stability with respect to two measures. The existence of smooth Liapunov functions for (4.4) is related to certain robustness properties.

4.1.3 Time varying systems

In [116], one of the authors proves a converse theorem for global asymptotic stability, by extending Kurzweil's proof to Filippov's solutions of equations with time dependent, discontinuous right hand side. In fact, using the same method as in [116] and [16, Lem. 4.1], it is possible to achieve a stronger result.

Theorem 4.4 *Assume that the origin is an equilibrium position for (4.1), and that F fulfills ($\mathbf{H_1}$), ($\mathbf{H_2}$), ($\mathbf{H_3}$) and ($\mathbf{H_4}$). If (4.1) is UGAS, then there exists a strict Liapunov function, which is locally Lipschitz continuous with respect to both variables.*

Surprisingly enough, the smoothness of the Liapunov function in Theorem 4.4 cannot be improved in general, as it is shown by next result (see [118, Prop. 1])

Proposition 4.3 *There exists a bounded Borel map $\lambda \colon \mathbb{R} \to \mathbb{R}$ such that the origin is UGAS for the system*

$$\dot{x} = \lambda(t)x \tag{4.5}$$

and such that there does not exist any strict (or weak) Liapunov function $V(t,x)$ in the large (or in the small) of class C^1.

Therefore, if we intend to construct a *smooth* Liapunov function, some additional assumption has to be made as far as the regularity of f with respect to time is concerned. Let us introduce the following

Definition 4.1 *Let $f \in \mathcal{L}^\infty_{\text{loc}}(\mathbb{R}^+ \times \mathbb{R}^n, \mathbb{R}^m)$, $m, n \geq 1$. The function f is said to be* essentially continuous with respect to time *(ECT in short) if there exists a set $\mathcal{N} \subset \mathbb{R}^+ \times \mathbb{R}^n$ of measure zero such that for each pair $(t_0, x_0) \in \mathbb{R}^+ \times \mathbb{R}^n$ and for each $\varepsilon > 0$, there exists $\delta > 0$ such that for each pair $(t, x) \in \mathbb{R}^+ \times \mathbb{R}^n$*

$$\left.\begin{array}{l} |t - t_0| + \|x - x_0\| < \delta \\ (t, x) \notin \mathcal{N} \\ (t_0, x) \notin \mathcal{N} \end{array}\right\} \Rightarrow \|f(t, x) - f(t_0, x)\| < \varepsilon. \quad (4.6)$$

Roughly speaking, the function f is ECT if the restriction of f to the complement of some set of measure zero is continuous with respect to time, in a (locally) uniform way with respect to x. Such a property is fulfilled when, for instance, f takes the form

$$f(t, x) := h(u(t), v(x)),$$

where $u : \mathbb{R}^+ \to \mathbb{R}^p$ is a (locally) Riemann integrable map, $v \in \mathcal{L}^\infty_{\text{loc}}(\mathbb{R}^n, \mathbb{R}^q)$ and $h \in C^0(\mathbb{R}^{p+q}, \mathbb{R}^m)$, $m, n, p, q \geq 1$. Indeed, a bounded function defined on a segment of $\mathbb{R}$ is Riemann integrable if, and only if, the set of points at which the function fails to be continuous is of measure zero [123]. Notice that every vector field f of the form

$$f(t, x) := f_0(x) + \sum_{i=1}^p u_i(t)\, f_i(x),$$

where for each $i \geq 0$, $f_i \in \mathcal{L}^\infty_{\text{loc}}(\mathbb{R}^n, \mathbb{R}^n)$ and for each $i \geq 1$, u_i is a piecewise continuous (or piecewise monotone) function, is of the mentioned form.

To state a converse Liapunov theorem which provides a *smooth* (i.e., C^∞) Liapunov function in a set valued context, we need to introduce a strengthened version of (**H**), namely the assumption

(**H′**): There exists a zero measure set $\mathcal{N}_0 \subset \mathbb{R}^+$ such that

(**H′₁**) For each $(t, x) \in (\mathbb{R}^+ \setminus \mathcal{N}_0) \times \mathbb{R}^n$, $F(t, x)$ is a nonempty convex compact set in $\mathbb{R}^n$;

(**H′₂**) For each $R > 0$, there exists a number $M > 0$ such that

$$(t \in [0, R] \setminus \mathcal{N}_0, \ \|x\| \leq R) \Rightarrow F(t, x) \subset \overline{B}_M;$$

124   Differential inclusions

($\mathbf{H'_3}$) The restriction $F|_{(\mathbb{R}^+ \setminus N_0) \times \mathbb{R}^n}$ is upper semicontinuous (u.s.c.), i.e., for each pair $(t_0, x_0) \in (\mathbb{R}^+ \setminus N_0) \times \mathbb{R}^n$ and each $\varepsilon > 0$ there exists $\delta > 0$ such that for any $(t, x) \in (\mathbb{R}^+ \setminus N_0) \times \mathbb{R}^n$

$$\|(t - t_0, x - x_0)\| < \delta \;\Rightarrow\; F(t, x) \subset F(t_0, x_0) + \overline{B}_\varepsilon.$$

The link between the essential continuity with respect to time and ($\mathbf{H'}$) appears clearly in next

Proposition 4.4 *Let $f \in \mathcal{L}^\infty_{\mathrm{loc}}(\mathbb{R}^+ \times \mathbb{R}^n, \mathbb{R}^n)$. If f is ECT, then ($\mathbf{H'}$) is fulfilled by $F(t, x) := \mathbf{K}_x f(t, x)$.*

We are now in a position to state the main result of this section, which contains Theorems 3.4, and Theorems 4.2 and 4.3.

Theorem 4.5 (CONVERSE OF SECOND LIAPUNOV THEOREM) *Assume that (4.1) is uniformly globally asymptotically stable (UGAS) at the origin and that ($\mathbf{H'}$) is fulfilled. Then for any $\lambda > 0$ there exist a function $V: \mathbb{R}^+ \times \mathbb{R}^n \to \mathbb{R}^+$ of class C^∞ and two functions a, b in $\mathcal{K}_0^\infty$ such that*

$$a(\|x\|) \leq V(t, x) \leq b(\|x\|) \quad \text{for all } t \geq 0 \text{ and } x \in \mathbb{R}^n; \tag{4.7}$$

$$\frac{\partial V}{\partial t}(t, x) + \langle \nabla_x V(t, x), v \rangle \leq -\lambda V(t, x)$$
$$\text{for all } t \in \mathbb{R}^+ \setminus N_0, \; x \in \mathbb{R}^n \text{ and } v \in F(t, x). \tag{4.8}$$

Furthermore, we may also require that V is time-periodic if F is time-periodic (with the same period), and that V is time-independent if F is time-independent.

Remark. We infer from (4.7) and (4.8) the standard estimate

$$\frac{\partial V}{\partial t}(t, x) + \langle \nabla_x V(t, x), v \rangle \leq -c(\|x\|),$$

where $c(\cdot)$ denotes some function in $\mathcal{K}_0^\infty$. (Take $c := \lambda a$.) We may also infer from (4.8) that V is exponentially decreasing along the trajectories of (4.1), meaning that for any trajectory $x(\cdot): [t_0, +\infty) \to \mathbb{R}^n$ of (4.1)

$$V(t, x(t)) \leq e^{-\lambda(t - t_0)} V(t_0, x(t_0)) \quad \forall t \geq t_0.$$

The interested reader is referred to next subsection for a detailed proof. Let us briefly outline this proof. In the first step, using an idea from [38], we show that the (uniform) global asymptotic stability also holds true for some perturbed system

$$\dot{x} \in F_2(t, x), \tag{4.9}$$

where $F(t,x) \subset F_2(t,x)$ for every x and a.e. t, and F_2 is locally Lipschitz continuous (for the Hausdorff distance) in $\mathbb{R}^+ \times (\mathbb{R}^n \setminus \{0\})$. The nice property we gain is that, for each pair (t_0, t_1) of times, the flow map $x_0 \mapsto \{x(t_1) : x(\cdot) \text{ solves (4.9) and } x(t_0) = x_0\}$ is locally Lipschitz continuous for the Hausdorff distance. In next step, we adapt some explicit construction by Yoshizawa in [159] of a locally Lipschitz continuous Liapunov function V_L to our differential inclusion setting. In the final step, we smooth V_L by a standard method using convolution with mollifiers and partition of unity.

4.1.4 Proof of the converse of second Liapunov theorem

The following set is introduced for notational convenience

$$E := ([-1, +\infty) \setminus N_0) \times \mathbb{R}^n.$$

Next a lemma, which is a special instance of Theorem 1.5, is reported here for the comfort of the reader.

Lemma 4.1 *Let $F = F(t,x)$ be a multivalued map fulfilling* ($\mathbf{H}_1$), *let $y : [T_1, T_2] \to \mathbb{R}^n$ denote some solution of (4.1) and let $b > 0$. Assume that F is Lipschitz continuous in the region $t \in [T_1, T_2]$, $\|x - y(t)\| \le b$. It means that there exists some constant $K > 0$ such that for any $t, \bar{t} \in [T_1, T_2]$ and any $x, \bar{x} \in \mathbb{R}^n$ with $\|x - y(t)\| \le b$, $\|\bar{x} - y(\bar{t})\| \le b$ we have*

$$h\big(F(t,x), F(\bar{t}, \bar{x})\big) \le K \, \|(t - \bar{t}, x - \bar{x})\|.$$

Let $(t_0, x_0) \in [T_1, T_2] \times \mathbb{R}^n$ be such that $\|x_0 - y(t_0)\| \le b$. Then there exists a solution $x(\cdot)$ of (4.1) such that $x(t_0) = x_0$ and

$$\|x(t) - y(t)\| \le \|x_0 - y(t_0)\| e^{K|t-t_0|} \tag{4.10}$$

as long as $\|x_0 - y(t_0)\| e^{K|t-t_0|} \le b$.

The construction of F_2, V_L and V will be first achieved in the general setting, i.e., when $F(t,x)$ is any time dependent set valued map. The main modifications to be brought in the time-periodic or time-independent cases will be indicated after.

FIRST STEP. (Regularization of F)

In order to facilitate the construction of a Liapunov function which is *smooth up to* $t = 0$, we extend F on $[-1, 0) \times \mathbb{R}^n$ by setting $F(t, x) = \{-x\}$. It is easily seen that F fulfills $(\mathbf{H'})$ on $[-1, +\infty) \times \mathbb{R}^n$ (provided that 0 is put into N_0) and that (4.1) remains UGAS $[-1, +\infty) \times \mathbb{R}^n$. To smooth F "from the inside" we need the following compactness lemma.

Lemma 4.2 *Assume that* $(\mathbf{H'})$ *holds true for* F *on* $[-1, +\infty) \times \mathbb{R}^n$. *Let* $R > 0$ *be given and let* $(t_1^j)_{j \geq 0}$, $(t_2^j)_{j \geq 0}$ *and* $(\delta^j)_{j \geq 0}$ *be sequences of numbers such that*

$$-1 \leq t_1^j \leq t_2^j \leq R \quad \text{for all } j,$$
$$\delta^j \to 0 \quad \text{as } j \to \infty.$$

Let $(x^j)_{j \geq 0}$ *be a sequence of absolutely continuous functions* $x^j : [t_1^j, t_2^j] \to \overline{B}_R$ *such that, for a.e.* $t \in [t_1^j, t_2^j]$

$$\dot{x}^j(t) \in \overline{co}\{F(\overline{B}_{\delta^j}(t, x^j(t)) \cap E)\}$$

$(\overline{B}_{\delta^j}(t, x^j(t))$ *denotes the closed ball for the sup-norm in* $\mathbb{R}^{n+1}$, *which is centered at* $(t, x^j(t))$ *and has* δ^j *as a radius.*) *Then there exist two numbers* $t_1, t_2 \in [-1, R]$, *a function* $x : [t_1, t_2] \to \overline{B}_R$ *and a sequence* $j^k \to \infty$ *such that* $x(\cdot)$ *is a solution of (4.1) on* $[t_1, t_2]$, $t_1^{j^k} \to t_1$, $t_2^{j^k} \to t_2$ *as* $k \to \infty$ *and*

$$\lim_{k \to \infty} x^{j^k}(t_1^{j^k}) = x(t_1), \quad \lim_{k \to \infty} x^{j^k}(t_2^{j^k}) = x(t_2).$$

Proof. In what follows, we sometimes need to take convergent subsequences of certain sequences of numbers or functions. In these cases, for the sake of simplicity and without loss of generality, we avoid using multiple indices: rather, we assume that the given sequence is actually convergent. Thus, since $-1 \leq t_1^j \leq t_2^j \leq R$, we may assume that $t_1^j \to t_1$ and $t_2^j \to t_2$. We may also assume that $\delta^j \leq 1$ for any $j \geq 0$. According to $(\mathbf{H'_2})$, there exists a constant $M > 0$ such that

$$t \in [-1, R+1] \setminus N_0, \ \|x\| \leq R+1 \Rightarrow F(t, x) \subset \overline{B}_M.$$

For $j \geq 0$ and $t \in [-1, R]$ we set

$$\tilde{x}^j(t) := \begin{cases} x^j(t_1^j) & \text{if} \quad -1 \leq t \leq t_1^j, \\ x^j(t) & \text{if} \quad t_1^j \leq t \leq t_2^j, \\ x^j(t_2^j) & \text{if} \quad t_2^j \leq t \leq R. \end{cases}$$

Since for any $j \geq 0$, $\|\tilde{x}^j(t)\| \leq R$ for every $t \in [-1, R]$ and $\|\dot{\tilde{x}}^j(t)\| \leq M$ for a.e. $t \in [-1, R]$, we infer that the sequence $(\tilde{x}^j)$ is bounded in the Sobolev space $H^1((-1, R), \mathbb{R}^n)$, hence a subsequence (also denoted by $(\tilde{x}^j)$) weakly converges in $H^1((-1, R), \mathbb{R}^n)$ towards a function x. It follows that $x^j \to x$ in $C^0([-1, R], \mathbb{R}^n)$ and that $\dot{\tilde{x}}^j \rightharpoonup \dot{x}$ in $L^2((-1, R), \mathbb{R}^n)$ (see e.g. [25, Thm. VIII.7]). As an easy consequence, $\tilde{x}^j(t_1^j) \to x(t_1)$ and $\tilde{x}^j(t_2^j) \to x(t_2)$. To prove that $x(\cdot)$ is indeed a solution of (4.1), i.e. that $\dot{x}(t) \in F(t, x(t))$ for a.e. $t \in [t_1, t_2]$, we consider the functional J defined on $L^2((-1, R), \mathbb{R}^n)$ by

$$J(w) := \int_{t_1}^{t_2} \mathrm{dist}\bigl(w(t), F(t, x(t))\bigr)\,dt.$$

Notice that the nonnegative map $t \mapsto \mathrm{dist}(w(t), F(t, x(t)))$ is *measurable*, according to [54, Prop. 3.4] [2]. It is easily seen that the functional J is well-defined, convex and continuous for the strong topology of $L^2((-1, R), \mathbb{R}^n)$, hence it is l.s.c. for the weak topology of $L^2((-1, R), \mathbb{R}^n)$ (see [25, Cor. III.8]). Since $\dot{\tilde{x}}^j \rightharpoonup \dot{x}$ in $L^2((-1, R), \mathbb{R}^n)$, we get

$$(0 \leq)\ J(\dot{x}) \leq \liminf_{j \to \infty} J(\dot{\tilde{x}}^j).$$

To prove $J(\dot{x}) = 0$, which is equivalent to (4.1) on (t_1, t_2), we only have to show that $\lim_{j \to \infty} J(\dot{\tilde{x}}^j) = 0$. Since $\mathrm{dist}\bigl(\dot{\tilde{x}}^j(t), F(t, x(t))\bigr) \leq 2M$ for a.e. $t \in (t_1, t_2)$ and each j, applying Lebesgue's theorem, we are done if we prove that for a.e. $t \in (t_1, t_2)$

$$\lim_{j \to \infty} \mathrm{dist}\bigl(\dot{\tilde{x}}^j(t), F(t, x(t))\bigr) = 0.$$

For a.e. $t_0 \in (t_1, t_2) \setminus N_0$, there exists an integer $j_0 \geq 0$ such that for any $j \geq j_0$, $t_0 \in (t_1^j, t_2^j)$ (hence the tilde may be dropped) and $\dot{x}^j(t_0)$ exists and it belongs to $\overline{\mathrm{co}}\{F(\overline{B}_{\delta^j}(t_0, x^j(t_0)) \cap E)\}$. Let $\varepsilon > 0$. According to ($\mathbf{H}_3'$), there exists $\delta > 0$ such that for any $(t, x) \in E$

$$\|(t - t_0, x - x(t_0))\| \leq \delta \Rightarrow F(t, x) \subset F(t_0, x(t_0)) + \overline{B}_\varepsilon.$$

By increasing j_0 if needed, we may also assume that for all $j \geq j_0$

$$\delta^j < \frac{\delta}{2} \quad \text{and} \quad \|x^j(t_0) - x(t_0)\| < \frac{\delta}{2}.$$

---

[2] To apply this result we need to check that the map $t \in (t_1, t_2) \mapsto F(t, x(t))$ is measurable. But the map $t \in (t_1, t_2) \setminus N_0 \mapsto (t, x(t)) \in E$ is continuous and $F_{|E}$ is u.s.c., hence the composite map $t \in (t_1, t_2) \setminus N_0 \mapsto F(t, x(t))$ is u.s.c., and therefore measurable. The same is true for the map $t \in (t_1, t_2) \mapsto F(t, x(t))$.

We infer that for all $j \geq j_0$, $\overline{B}_{\delta^j}(t_0, x^j(t_0)) \subset \overline{B}_\delta(t_0, x(t_0))$ and

$$\overline{\mathrm{co}}\{F(\overline{B}_{\delta^j}(t_0, x^j(t_0)) \cap E)\} \subset F(t_0, x(t_0)) + \overline{B}_\varepsilon.$$

It follows that $\mathrm{dist}(\dot{\tilde{x}}^j(t_0), F(t_0, x(t_0))) \leq \varepsilon$ for $j \geq j_0$, as required. ∎

In what follows, $\delta = \delta(t, x)$ denotes any continuous function $[-1, +\infty) \times \mathbb{R}^n \to \mathbb{R}^+$ such that

$$\forall (t, x) \in [-1, +\infty) \times \mathbb{R}^n \quad \delta(t, x) = 0 \iff x = 0.$$

If such a function δ is given, we set for any $(t, x) \in [-1, +\infty) \times \mathbb{R}^n$

$$F_1(t, x) := \overline{\mathrm{co}}\{F(\overline{B}_{\delta(t,x)}(t, x) \cap E)\}. \tag{4.11}$$

We aim to prove that, for a suitable choice of δ, the system

$$\dot{x} \in F_1(t, x), \quad t \geq -1, \; x \in \mathbb{R}^n \tag{4.12}$$

inherits the (uniform) global asymptotic stability of (4.1). Before proving such a property, let us observe that, for any choice of δ, the multivalued map F_1 fulfills (**H′**) on $[-1, +\infty) \times \mathbb{R}^n$, as well. Indeed,

$$F_1(t, 0) = \begin{cases} F(t, 0) & \text{if } t \notin N_0, \\ \emptyset & \text{if } t \in N_0, \end{cases}$$

hence (**H′₁**) and (**H′₂**) are obvious. Next (**H′₃**) follows easily from the upper-semicontinuity of the multivalued maps $F_{|E}$ and $(t, x) \mapsto \overline{B}_{\delta(t,x)}(t, x) \cap E$ ($\in 2^E$).

Since (4.1) is UGAS, there exists a sequence $(\varphi_i)_{i \in \mathbb{Z}}$ of continuous decreasing functions $[0, +\infty) \to (0, +\infty)$ fulfilling the following properties:
(i) For any pair $(t_0, x_0) \in [-1, +\infty) \times \mathbb{R}^n$ and any solution $x(\cdot)$ of (4.1) such that $x(t_0) = x_0$,

$$\|x_0\| \leq 2^i \implies \|x(t_0 + h)\| < \varphi_i(h) \text{ for all } h \geq 0;$$

(ii) For each $i \in \mathbb{Z}$, $\varphi_i(t) \to 0$ as $t \to \infty$;
(iii) $(\varphi_i(0))_{i \in \mathbb{Z}}$ is a nondecreasing sequence such that $\varphi_i(0) \to 0$ as $i \to -\infty$ and $\varphi_i(0) \to +\infty$ as $i \to +\infty$.
Indeed, it is sufficient to set $\varphi_i(h) := \beta(h, 2^i)$, where β is given by Proposition 4.1. Notice that $\varphi_i(0) > 2^i$. For each $i \in \mathbb{Z}$, let $p_i \in \mathbb{Z}$ be the greatest integer such that $\varphi_{p_i}(0) \leq 2^{i-1}$. Clearly, $p_i < i - 1$, the sequence (p_i) is nondecreasing and $p_i \to \pm\infty$ as $i \to \pm\infty$.

Remark 4.1 *As a direct consequence, any compact set in $\mathbb{R}^n \setminus \{0\}$ intersects the sets $\{x\colon 2^{p_i} \leq \|x\| \leq \varphi_i(0)\}$ for at most a finite number of i.*

For each $i \in \mathbb{Z}$, pick a number $T_i \geq 1$ such that $\varphi_i(T_i) \leq 2^{i-1}$ and set $\hat{T}_i = \max(T_i, \max(T_j : p_j = i))$ when there exists $j \in \mathbb{Z}$ s.t. $p_j = i$, $\hat{T}_i = T_i$ otherwise.

Lemma 4.3 *Let $i \in \mathbb{Z}$ and $k \in \{-1\} \cup \mathbb{N}$. Then there exists a constant $\delta > 0$ such that, for any solution $x(\cdot)$ of*

$$\dot{x} \in \overline{\mathrm{co}}\{F(\overline{B}_\delta(t, x) \cap E)\} \tag{4.13}$$

fulfilling $\|x(t_0)\| \leq 2^i$ for some time $t_0 \in [k, k+1]$, we have

$$\|x(t_0 + h)\| < \varphi_i(h) \quad \forall h \in [0, \hat{T}_i].$$

Proof. If the statement of Lemma 4.3 is false, then there exists a sequence of positive numbers $\delta^j \searrow 0$ and a sequence of absolutely continuous functions $(x^j)_{j \geq 0}$, with $x^j : [t_0^j, t_1^j] \to \mathbb{R}^n$, $t_0^j \in [k, k+1]$, $t_0^j \leq t_1^j \leq t_0^j + \hat{T}_i$, such that

$$\dot{x}^j(t) \in \overline{\mathrm{co}}\{F(\overline{B}_{\delta^j}(t, x^j(t)) \cap E)\} \quad \text{for a.e. } t \in [t_0^j, t_1^j],$$
$$\|x^j(t_0^j)\| \leq 2^i, \ \|x^j(t)\| < \varphi_i(t - t_0^j) \text{ for all } t \in [t_0^j, t_1^j), \text{ and}$$
$$\|x^j(t_1^j)\| = \varphi_i(t_1^j - t_0^j).$$

Using Lemma 4.2 and extracting subsequences if needed, we may also assume that for some numbers $t_0, t_1 \in [k, k+1+\hat{T}_i]$ and some solution $x(\cdot) : [t_0, t_1] \to \mathbb{R}^n$ of (4.1), we have $(t_0^j, x^j(t_0^j)) \to (t_0, x(t_0))$ and $(t_1^j, x^j(t_1^j)) \to (t_1, x(t_1))$ as $j \to \infty$. Hence $\|x(t_0)\| \leq 2^i$ and $\|x(t_1)\| = \varphi_i(t_1 - t_0)$, which contradicts the definition of φ_i. ∎

For each pair $(i, k) \in \mathbb{Z} \times (\{-1\} \cup \mathbb{N})$, we denote by δ_i^k the number $\delta > 0$ given by Lemma 4.3. Pick any continuous function $\delta : [-1, +\infty) \times \mathbb{R}^n \to \mathbb{R}^+$ fulfilling

$$\delta(t, x) = 0 \iff x = 0 \quad \text{and} \tag{4.14}$$

$$\delta(t, x) < \min(\delta_i^k, \delta_{p_i}^k) \quad \text{whenever } k \leq t \text{ and } 2^{p_i} \leq \|x\| \leq \varphi_i(0). \tag{4.15}$$

Notice that such a function δ exists, due to Remark 4.1. Replacing (if needed) δ by the inf-convolution function

$$\tilde{\delta}(t, x) = \inf_{(s,y) \in [-1,+\infty) \times \mathbb{R}^n} \big(\delta(s, y) + \|(t-s, x-y)\|\big),$$

we may also require that δ be Lipschitz continuous, with 1 as a Lipschitz constant[3]. To prove the (uniform) global asymptotic stability of (4.12) we need the

Lemma 4.4 *Let $\delta(\cdot)$ fulfill (4.14)-(4.15), and let $x(\cdot)$ be any solution of (4.12). Assume that $\|x(t_0)\| \leq 2^i$ for some time $t_0 \geq -1$ and some $i \in \mathbb{Z}$. Then*

(a) $\quad \|x(t_0 + h)\| < \varphi_i(0) \quad \forall h \in [0, T_i],$
(b) $\quad \|x(t_0 + T_i)\| \leq 2^{i-1}.$

Proof. If (a) is false, then there exist two times t_1, t_2 such that $t_0 \leq t_1 < t_2 \leq t_0 + T_i$ and $2^i = \|x(t_1)\| < \|x(t)\| < \|x(t_2)\| = \varphi_i(0)$ for $t_1 < t < t_2$. Set $k := [t_1]$. Since $2^{p_i} \leq \|x(t)\| \leq \varphi_i(0)$ for $t \in [t_1, t_2]$, we infer from (4.15) that $x(\cdot)$ is also a solution of

$$\dot{x} \in \overline{\mathrm{co}}\{F(\overline{B}_{\delta_i^k}(t, x) \cap E)\} \tag{4.16}$$

on $[t_1, t_2]$, hence, by Lemma 4.3, $\|x(t_2)\| < \varphi_i(t_2 - t_1) < \varphi_i(0)$, which contradicts some property of t_2. We now turn to the proof of (b). Assume that $\|x(t_0 + T_i)\| > 2^{i-1}$. If $2^{p_i} \leq \|x(t_0 + h)\|$ ($\leq \varphi_i(0)$) for all $h \in [0, T_i]$, then $x(\cdot)$ again solves (4.16) on $[t_0, t_0 + T_i]$ (with here $k := [t_0]$), hence, by Lemma 4.3, $\|x(t_0 + T_i)\| < \varphi_i(T_i) \leq 2^{i-1}$, leading to a contradiction. We infer that there exist two times t_1, t_2 such that $t_0 \leq t_1 < t_2 \leq t_0 + T_i$ and $2^{p_i} = \|x(t_1)\| < \|x(t)\| < \|x(t_2)\| = 2^{i-1}$ for $t_1 < t < t_2$. Set again $k := [t_1]$. Since $\delta(t, x(t)) < \delta_{p_i}^k$ for $t_1 \leq t \leq t_2$ and $t_2 - t_1 \leq T_i \leq \hat{T}_{p_i}$, we infer from Lemma 4.3 (applied with p_i instead of i) that $\|x(t_2)\| < \varphi_{p_i}(0) \leq 2^{i-1}$, contradicting some property of t_2. ∎

Corollary 4.1 *Let δ and F_1 be as above. Then (4.12) is UGAS on $[-1, +\infty) \times \mathbb{R}^n$.*

Indeed, we infer from (a) and (b) that $\|x(t_0 + h)\| < \varphi_{i-l}(0)$ for all $l \in \mathbb{N}$ and all $h \geq \sum_{j=i-l+1}^{i} T_j$.

We are now in position to define the enlarged (and regularized) system $\dot{x} \in F_2(t, x)$. We need some suitable partition of unity. We set

$$U := (-1, +\infty) \times (\mathbb{R}^n \setminus \{0\})$$

---

[3]Indeed, $\tilde{\delta}(t, 0) = 0 < \tilde{\delta}(t, x) \leq \delta(t, x)$ for $t \geq -1$, $x \neq 0$, and it is easy to see that $\|\tilde{\delta}(t, x) - \tilde{\delta}(s, y)\| \leq \|(t - s, x - y)\|$ for all pairs $(t, x), (s, y) \in [-1, +\infty) \times \mathbb{R}^n$.

and, for any $(t, x) \in U$,

$$W(t,x) := \{(s,y) \in U : \|(s-t, y-x)\| < \frac{1}{3}\delta(t,x)\}. \tag{4.17}$$

We have the following

CLAIM 1. *The family $\bigl(W(t,x)\bigr)_{(t,x)\in U\cap E}$ is an open covering of U.*

To prove the claim, we have to check that for any $(t_0, x_0) \in U$ there exists some $(t, x) \in U \cap E$ such that $(t_0, x_0) \in W(t, x)$. If $(t_0, x_0) \in U \cap E = \bigl((-1, +\infty) \setminus N_0\bigr) \times (\mathbb{R}^n \setminus \{0\})$, then $(t_0, x_0) \in W(t_0, x_0)$ (since $\delta(t_0, x_0) > 0$). If $(t_0, x_0) \in N_0 \times (\mathbb{R}^n \setminus \{0\})$, since N_0, as a set of measure zero, has a dense complement in $(-1, +\infty)$, we may find a time $t \in (-1, +\infty) \setminus N_0$ such that $|t - t_0| < \frac{1}{6}\delta(t_0, x_0)$ and $\delta(t, x_0) > \frac{1}{2}\delta(t_0, x_0)$. Then $(t_0, x_0) \in W(t, x_0)$. ∎

Let $(\psi_i)_{i\geq 1}$ be some C^∞ partition of unity on U subordinate to the open covering $\bigl(W(t,x)\bigr)_{(t,x)\in U\cap E}$ of U. It means that 1) each ψ_i is a nonnegative function of class C^∞ on $\mathbb{R}^{n+1}$, with its support contained in $W(t_i, x_i)$ for some $(t_i, x_i) \in U \cap E$, 2) $\sum_{i=1}^{\infty} \psi_i(t, x) = 1$ for any $(t, x) \in U$ and 3) for each $(t, x) \in U$, there exists a number $\rho > 0$ such that $\psi_i \equiv 0$ on $\overline{B}_\rho(t, x)$ for all $i \geq 1$ except a *finite* number.

Set for any $(t, x) \in (-1, +\infty) \times \mathbb{R}^n$

$$F_2(t,x) := \begin{cases} \sum_{i=1}^{\infty} \psi_i(t,x)\, \overline{\mathrm{co}}\,\{F\bigl(\overline{B}_{\frac{1}{3}\delta(t_i,x_i)}(t_i,x_i) \cap E\bigr)\} & \text{if } x \neq 0, \\ F(t,0) & \text{if } x = 0. \end{cases} \tag{4.18}$$

Since on compact subsets of U the sum in (4.18) is *finite*, we infer that F_2 is locally Lipschitz continuous (for the Hausdorff distance) on U. The links between F, F_1 and F_2 appear clearly in next claims.

CLAIM 2. $\forall t \in (-1, +\infty) \setminus N_0$, $\forall x \in \mathbb{R}^n$, $F(t, x) \subset F_2(t, x)$.

The result is obvious for $x = 0$, so let $x \neq 0$ and $t \in (-1, +\infty) \setminus N_0$. Let $i \geq 1$ be such that $\psi_i(t, x) > 0$, hence $(t, x) \in W(t_i, x_i)$. By (4.17), $\|(t - t_i, x - x_i)\| < \frac{1}{3}\delta(t_i, x_i)$, hence

$$F(t,x) \subset F\bigl(\overline{B}_{\frac{1}{3}\delta(t_i,x_i)}(t_i,x_i) \cap E\bigr),$$

from which we infer that $F(t, x) \subset F_2(t, x)$. ∎

CLAIM 3. $\forall (t,x) \in U \quad F_2(t,x) \subset F_1(t,x)$.

To prove this claim, consider any pair $(t,x) \in U$ and let $i \geq 1$ be such that $\psi_i(t,x) > 0$. Since δ is Lipschitz continuous with 1 as a Lipschitz constant, we have

$$\delta(t_i, x_i) - \delta(t,x) \leq \|(t-t_i, x-x_i)\| < \frac{1}{3}\delta(t_i, x_i) \tag{4.19}$$

hence

$$\overline{B}_{\frac{1}{3}\delta(t_i,x_i)}(t_i, x_i) \subset \overline{B}_{\frac{2}{3}\delta(t_i,x_i)}(t,x) \subset \overline{B}_{\delta(t,x)}(t,x),$$

and we infer that $F_2(t,x) \subset \overline{\mathrm{co}}\{F(\overline{B}_{\delta(t,x)}(t,x) \cap E)\} = F_1(t,x)$. ∎

Remark 4.2 (**H**$'$) is also fulfilled by F_2 on $(-1, +\infty) \times \mathbb{R}^n$, with the same N_0 as for F. Indeed (**H**$'_1$) is clear, (**H**$'_2$) follows from Claim 3, and (**H**$'_3$) is obvious on U, since F_2 is locally Lipschitz continuous on U. Finally, using the continuity of δ, (4.14), (4.19), and the fact that $F_{|E}$ is u.s.c., we easily check that $F_{2|E}$ is u.s.c. at each $(t_0, 0)$ for $t_0 \in (-1, +\infty) \setminus N_0$.

Corollary 4.2 *The origin is UGAS on $(-1, +\infty) \times \mathbb{R}^n$ for the system*

$$\dot{x} \in F_2(t,x) \tag{4.20}$$

Corollary 4.2 is a direct consequence of Corollary 4.1 and Claim 3. This completes the first step in the proof of Theorem 4.5. We now proceed to the

$\boxed{\text{Second step.}}$ (Construction of a locally Lipschitz continuous strict Liapunov function)

For any $(t_0, x_0) \in (-1, +\infty) \times \mathbb{R}^n$, let $\mathcal{S}_{t_0, x_0}$ denote the set of solutions $x(\cdot) \colon [t_0, +\infty)$ of the Cauchy problem

$$\begin{cases} \dot{x} \in F_2(t,x) \\ x(t_0) = x_0. \end{cases} \tag{4.21}$$

Set for any $q \in \mathbb{N}^*$ and any $r \in [0, +\infty)$,

$$G_q(r) := \max\left(0, r - \frac{1}{q}\right). \tag{4.22}$$

Finally, set for any $q \in \mathbb{N}^*$ and any $(t,x) \in (-1, +\infty) \times \mathbb{R}^n$,

$$V_q(t,x) := \sup_{\varphi \in \mathcal{S}_{t,x}} \sup_{\tau \geq 0} e^{2\lambda \tau} G_q(\|\varphi(t+\tau)\|), \tag{4.23}$$

where $\lambda > 0$ is the number appearing in the statement of Theorem 4.5. The system (4.20) being UGAS, for any $R > 0$ and any $q \in \mathbb{N}^*$ there exist positive numbers $m(R)$ and $T(R,q)$ such that for each pair $(t_0, x_0) \in (-1, +\infty) \times \mathbb{R}^n$ and for each solution $\varphi \in \mathcal{S}_{t_0, x_0}$, if $\|x_0\| \leq R$ then $\|\varphi(t_0 + \tau)\| < m(R)$ for all $\tau \geq 0$ and $\|\varphi(t_0 + \tau)\| < \frac{1}{q}$ for all $\tau > T(R, q)$. Furthermore, $T(R, q)$ and $m(R)$ may be chosen in such a way that $T(\cdot, q)$ is nondecreasing for each $q \geq 1$ and $m(\cdot)$ is in $\mathcal{K}_0^\infty$. As a direct consequence, we obtain the

Lemma 4.5 Let $R > 0$ and $(t, x) \in (-1, +\infty) \times \overline{B}_R$. Then, for each $q \in \mathbb{N}^*$,

$$\begin{aligned} G_q(\|x\|) &\leq V_q(t, x) \\ &= \sup_{\varphi \in \mathcal{S}_{t,x}} \sup_{\tau \in [0, T(R,q)]} e^{2\lambda \tau} G_q(\|\varphi(t + \tau)\|) \\ &\leq e^{2\lambda T(R,q)} m(R) < +\infty \, . \end{aligned} \quad (4.24)$$

Indeed, for any $\varphi \in \mathcal{S}_{(t,x)}$ and any $\tau > T(R, q)$, $\|\varphi(t+\tau)\| < \frac{1}{q}$, hence $G_q(\|\varphi(t+\tau)\|) = 0$, which implies the equality in (4.24). Next, we remark that $G_q(\|\varphi(t+\tau)\|) \leq \|\varphi(t+\tau)\| \leq m(R)$. Finally, the first inequality is obvious (consider $\tau = 0$ in (4.23)). ∎

Notice that $V_q(t, 0) = 0$ for any $t > -1$, since $\mathcal{S}_{t,0}$ contains only the trivial solution $x \equiv 0$. Another important property of V_q, namely the local Lipschitz continuity, is stated in next

Proposition 4.5 Let $q \in \mathbb{N}^*$ and $R > 0$ be given. Then there exists a positive constant $C_q(R)$ such that for any $t_1, t_2 \in [-\frac{R}{R+1}, R]$ and any $x_1, x_2 \in \overline{B}_R$,

$$|V_q(t_1, x_1) - V_q(t_2, x_2)| \leq C_q(R) \|(t_1 - t_2, x_1 - x_2)\| \, . \quad (4.25)$$

The proof of Proposition 4.5 is somewhat long and technical. It will be reported at the end of Step 2. Without loss of generality, we may assume that for each $q \in \mathbb{N}^*$ the function $R > 0 \mapsto C_q(R) > 0$ is nondecreasing. We are now in position to define a locally Lipschitz continuous (strict) Liapunov function for (4.20). For any $(t, x) \in (-1, +\infty) \times \mathbb{R}^n$ we set

$$V_L(t, x) := \sum_{q=1}^{+\infty} \frac{2^{-q}}{1 + C_q(q)} e^{-2\lambda T(q,q)} V_q(t, x). \quad (4.26)$$

CLAIM 4. There exists a function $a_L(\cdot)$ in $\mathcal{K}_0^\infty$ such that

$$a_L(\|x\|) \leq V_L(t, x) \quad \forall (t, x) \in (-1, +\infty) \times \mathbb{R}^n. \quad (4.27)$$

To prove the claim, set for any $r \geq 0$

$$a_L(r) := \sum_{q=1}^{+\infty} \frac{2^{-q}e^{-2\lambda T(q,q)}}{1+C_q(q)} G_q(r). \qquad (4.28)$$

Clearly, $a_L(\cdot)$ is a well-defined, increasing, Lipschitz continuous function (with $\sum_{q=1}^{+\infty} \frac{2^{-q}e^{-2\lambda T(q,q)}}{1+C_q(q)}$ as a Lipschitz constant) such that $\lim_{r\to+\infty} a_L(r) = +\infty$. (4.27) follows from (4.24) and (4.26). ∎

CLAIM 5. Let $R > 0$ be given. Then there exists a constant $L(R) > 0$ such that for any $t_1, t_2 \in \left[-\frac{R}{R+1}, R\right]$ and any $x_1, x_2 \in \overline{B}_R$

$$|V_L(t_1, x_1) - V_L(t_2, x_2)| \leq L(R) \|(t_1 - t_2, x_1 - x_2)\|. \qquad (4.29)$$

Indeed,

$$\begin{aligned} V_L(t_1, x_1) &\leq V_L(t_2, x_2) + \sum_{q=1}^{+\infty} \frac{2^{-q}e^{-2\lambda T(q,q)}}{1+C_q(q)} |V_q(t_1, x_1) - V_q(t_2, x_2)| \\ &\leq V_L(t_2, x_2) + \left(\sum_{q=1}^{+\infty} 2^{-q} \frac{C_q(R)}{1+C_q(q)} e^{-2\lambda T(q,q)}\right) \|(t_1 - t_2, x_1 - x_2)\| \end{aligned} \qquad (4.30)$$

due to Proposition 4.5. For any $R > 0$, set

$$L(R) := \sum_{q=1}^{+\infty} 2^{-q} \frac{C_q(R)}{1+C_q(q)} e^{-2\lambda T(q,q)}. \qquad (4.31)$$

Let us prove that $L(R) < +\infty$ for any $R > 0$. Using the fact that $C_q(q) \geq C_q(R)$ if $q > R$, we get

$$L(R) \leq \sum_{q=1}^{[R]} 2^{-q} \frac{C_q(R)}{1+C_q(q)} e^{-2\lambda T(q,q)} + \sum_{q=[R]+1}^{+\infty} 2^{-q} < +\infty.$$

Notice also that $L(\cdot)$ is a nondecreasing function. If $x_2 = 0$, then $V_L(t_2, x_2) = 0$ and it follows from (4.30) (applied with $t_2 = t_1$) that $V_L(t_1, x_1) \leq L(R) \|x_1\| < +\infty$, hence V_L is well-defined. Exchanging the roles of (t_1, x_1) and (t_2, x_2) in (4.30), we get (4.29). ∎

CLAIM 6. There exists a function $b_L(\cdot)$ in $\mathcal{K}_0^\infty$ such that

$$V_L(t, x) \leq b_L(\|x\|) \quad \forall (t, x) \in (-1, +\infty) \times \mathbb{R}^n. \qquad (4.32)$$

Indeed, it follows from (4.24) that for any $R > 0$ and any $(t, x) \in (-1, +\infty) \times \overline{B}_R$

$$\begin{aligned} V_L(t, x) &\leq \sum_{q=1}^{+\infty} 2^{-q} \frac{e^{2\lambda\left(T(R,q) - T(q,q)\right)}}{1+C_q(q)} m(R) \\ &\leq \left[\sum_{q=1}^{[R]} 2^{-q} \frac{e^{2\lambda\left(T(R,q) - T(q,q)\right)}}{1+C_q(q)} + 1\right] m(R) =: \tilde{m}(R). \end{aligned}$$

(We used the fact that $T(R,q) \leq T(q,q)$ for $q \geq [R]+1$.) Since $\tilde{m}(\cdot)$ is nondecreasing and $\lim_{R\to 0^+} \tilde{m}(R) = 0$, there exists a function $b_L(\cdot)$ in $\mathcal{K}_0^\infty$ such that $\tilde{m}(R) \leq b(R)$ for all $R > 0$. Then $V_L(t,x) \leq b_L(\|x\|)$, as wanted. ∎

The last useful property of V_L we report here is the way V_L decreases along the trajectories of (4.20).

CLAIM 7. Let $(t_0, x_0) \in (-1, +\infty) \times \mathbb{R}^n$ and let $\psi \in \mathcal{S}_{t_0, x_0}$. Then

$$V_L(t_0 + h, \psi(t_0 + h)) \leq e^{-2\lambda h} V_L(t_0, x_0) \quad \forall h \geq 0. \tag{4.33}$$

This claim is a direct consequence of next result, which will be needed when proving Proposition 4.5.

Lemma 4.6 Let $(t_0, x_0) \in (-1, +\infty) \times \mathbb{R}^n$ and let $\psi \in \mathcal{S}_{t_0, x_0}$. Then, for any $q \in \mathbb{N}^*$,

$$V_q(t_0 + h, \psi(t_0 + h)) \leq e^{-2\lambda h} V_q(t_0, x_0) \quad \forall h \geq 0. \tag{4.34}$$

To prove the lemma, consider for any $\varphi \in \mathcal{S}_{t_0+h, \psi(t_0+h)}$ the function $\overline{\varphi} : [t_0, +\infty) \to \mathbb{R}^n$ defined by

$$\overline{\varphi}(t) := \begin{cases} \psi(t) & \text{for } t_0 \leq t \leq t_0 + h \\ \varphi(t) & \text{for } t_0 + h \leq t. \end{cases}$$

Clearly $\overline{\varphi} \in \mathcal{S}_{t_0, x_0}$, hence

$$\begin{aligned} V_q(t_0, x_0) &\geq \sup_{\tau \geq 0} e^{2\lambda \tau} G_q(\|\overline{\varphi}(t_0 + \tau)\|) \\ &\geq e^{2\lambda h} \sup_{\tau \geq 0} e^{2\lambda \tau} G_q(\|\varphi(t_0 + h + \tau)\|). \end{aligned}$$

$\varphi \in \mathcal{S}_{t_0+h, \psi(t_0+h)}$ being arbitrary, we infer that $V_q(t_0, x_0) \geq e^{2\lambda h} V_q(t_0 + h, \psi(t_0 + h))$. ∎

Corollary 4.3 For a.e. $(t_0, x_0) \in U$,

$$\forall v \in F_2(t_0, x_0), \quad \frac{\partial V_L}{\partial t}(t_0, x_0) + \langle \nabla_x V_L(t_0, x_0), v \rangle \leq -2\lambda V_L(t_0, x_0).$$

Proof. Since V_L is locally Lipschitz continuous on U (by Proposition 4.5), we infer from Rademacher's theorem that V_L is differentiable at (t_0, x_0) for almost every $(t_0, x_0) \in U$. Therefore, it is sufficient to prove that for *every* pair $(t_0, x_0) \in U$,

$$\forall v \in F_2(t_0, x_0), \quad \limsup_{h \to 0^+} \frac{V_L(t_0 + h, x_0 + hv) - V_L(t_0, x_0)}{h} \leq -2\lambda V_L(t_0, x_0).$$

136   Differential inclusions

Notice first that for each $v \in F_2(t_0, x_0)$, there exists a solution $x(t)$ of (4.20) such that $x(t_0) = x_0$ and $\dot{x}(t_0) = v$. Indeed, we may infer from the local Lipschitz continuity of F_2 that the map

$$g(s,y) := \pi_{F_2(s,y)}(v), \quad (s,y) \in U, \tag{4.35}$$

(where $\pi_{F_2(s,y)}$ denotes the projection on the convex compact set $F_2(s,y)$) is *continuous*. Hence, by Peano's theorem, there exists a solution $x(t)$ on some interval $[t_0, t_0 + \varepsilon]$ of the (classical) Cauchy problem

$$\begin{cases} \dot{x} = g(t,x) \; (\in F_2(t,x)) \\ x(t_0) = x_0. \end{cases}$$

Obviously, $\dot{x}(t_0) = g(t_0, x_0) = v$. Now there exists a constant $K > 0$ such that for any (t_1, x_1), (t_2, x_2) in some neighborhood of (t_0, x_0),

$$|V_L(t_1, x_1) - V_L(t_2, x_2)| \leq K \, \|(t_1 - t_2, x_1 - x_2)\|.$$

It follows that for $h > 0$ small enough

$$\frac{V_L(t_0 + h, x_0 + hv) - V_L(t_0, x_0)}{h}$$
$$= \frac{V_L(t_0 + h, x_0 + hv) - V_L(t_0 + h, x(t_0 + h))}{h}$$
$$+ \frac{V_L(t_0 + h, x(t_0 + h)) - V_L(t_0, x_0)}{h}$$
$$\leq K \left\| \frac{x(t_0 + h) - x_0}{h} - v \right\| + \frac{e^{-2\lambda h} - 1}{h} V_L(t_0, x_0),$$

hence

$$\limsup_{h \to 0^+} \frac{V_L(t_0 + h, x_0 + hv) - V_L(t_0, x_0)}{h} \leq -2\lambda V_L(t_0, x_0).$$

∎

It remains to prove Proposition 4.5. We first need the following elementary lemma.

Lemma 4.7 *Let V be a function defined on a set $K := \prod_{i=1}^n [a_i, b_i] \subset \mathbb{R}^n$. Assume the existence of a constant $L > 0$ such that, for any $x_0 \in K$ we may find a number $\eta_0 > 0$ for which*

$$|V(x) - V(x_0)| \leq L \, \|x - x_0\| \quad \forall x \in \overline{B}_{\eta_0}(x_0) \cap K. \tag{4.36}$$

Then V is Lipschitz continuous on K, with nL as a Lipschitz constant.

Remark 4.3 *(4.36) is weaker than the assumption $|V(x) - V(y)| \leq L\,\|x - y\| \quad \forall x, y \in \overline{B}_{\eta_0}(x_0) \cap K$, for which Lemma 4.7 is classical.*

Proof. Let $x^1 = (x_1^1, \ldots, x_n^1)$ and $x^2 = (x_1^2, \ldots, x_n^2)$ be given in K. Then

$$|V(x^1) - V(x^2)| \leq \sum_{j=1}^n |V(x_1^1, \ldots, x_j^1, x_{j+1}^2, \ldots, x_n^2) - V(x_1^1, \ldots, x_{j-1}^1, x_j^2, \ldots, x_n^2)|$$

so we are done if we prove that for any $j \in [1, n]$

$$|V(x_1^1, \ldots, x_j^1, x_{j+1}^2, \ldots, x_n^2) - V(x_1^1, \ldots, x_{j-1}^1, x_j^2, \ldots, x_n^2)| \leq L|x_j^1 - x_j^2|.$$

Fix $j \in [1, n]$ and assume for instance that $x_j^1 \leq x_j^2$. Define a function $f : [x_j^1, x_j^2] \to \mathbb{R}$ by $f(s) = V(x_1^1, \ldots, x_{j-1}^1, s, x_{j+1}^2, \ldots, x_n^2)$. Then it follows from (4.36) that $|\overline{D^+} f(s)| \leq L$ for all $s \in [x_j^1, x_j^2)$, hence $|f(x_j^1) - f(x_j^2)| \leq L|x_j^1 - x_j^2|$, as wanted. ∎

Proposition 4.5 is a direct consequence of Lemma 4.7 and of the following result.

Proposition 4.6 *Let $q \in \mathbb{N}^*$ and $R > 0$ be given. Then there exists a positive constant $L > 0$ such that for any $t_0 \in [-\frac{R}{1+R}, R]$ and any $x_0 \in \overline{B}_R$, we may find a number $\eta_0 > 0$ for which*

$$|V_q(t, x) - V_q(t_0, x_0)| \leq L \|(t - t_0, x - x_0)\| \quad \forall (t, x) \in \overline{B}_{\eta_0}(t_0, x_0). \tag{4.37}$$

Proof of Proposition 4.6. We need to introduce some numbers depending only on R and q. Let $T = T(R+1, q)$. By local Lipschitz continuity of F_2 on U, there exists a number $K > 0$ such that for any pairs $(t_1, x_1), (t_2, x_2)$ with $-\frac{R+1}{R+2} \leq t_i \leq R+T+1$ and $\frac{1}{2}m^{-1}(\frac{1}{q}) \leq \|x_i\| \leq m(R+2)$ for $i = 1, 2$, we have

$$h(F_2(t_1, x_1), F_2(t_2, x_2)) \leq K\|(t_1 - t_2, x_1 - x_2)\|. \tag{4.38}$$

Let $M \geq 1$ be such that for any $(t, x) \in \left([-\frac{R+1}{R+2}, R+T+1] \setminus N_0\right) \times \overline{B}_{m(R+2)}$, we have

$$F_2(t, x) \subset \overline{B}_M. \tag{4.39}$$

We set

$$L := e^{2\lambda T}\left((M+1)e^{K(T+1)} + 2\lambda m(R+1)\right). \tag{4.40}$$

Pick a number $\overline{\eta}_0$ such that

$$0 < \overline{\eta}_0 < \min\left\{\frac{R+1}{R+2} - \frac{R}{R+1},\right.$$
$$\left. e^{-K(T+1)} \min\left\{m(R+2) - m(R+1), \frac{1}{2}m^{-1}\left(\frac{1}{q}\right)\right\}\right\} \tag{4.41}$$

and set

$$b := \overline{\eta}_0 e^{K(T+1)}. \tag{4.42}$$

We begin with the

138   Differential inclusions

Lemma 4.8 *Let* $(t_0, x_0) \in [-\frac{R}{R+1}, R] \times \overline{B}_R$ *and* $(t_1, x_1) \in \overline{B}_{\overline{\eta}_0}(t_0, x_0)$. *If* $V_q(t_1, x_1) > 0$ *then for any* $\varphi_1 \in \mathcal{S}_{t_1, x_1}$ *fulfilling* $\|\varphi_1(t_1 + \tau)\| > \frac{1}{q}$ *for some* $\tau \in [0, T]$ *(such functions exist by (4.22) and (4.24)) we have*

$$m^{-1}\left(\frac{1}{q}\right) < \|\varphi_1(t_1 + h)\| < m(R+1) \quad \text{for } 0 \leq h \leq \tau. \tag{4.43}$$

Furthermore, the region $t_1 \leq t \leq t_1 + \tau$, $\|x - \varphi_1(t)\| \leq b$ *is contained in the region* $-\frac{R+1}{R+2} \leq t \leq R+T+1$, $\frac{1}{2}m^{-1}(\frac{1}{q}) \leq \|x\| \leq m(R+2)$, *where* F_2 *is Lipschitz continuous with constant* K.

Proof. Since $\overline{\eta}_0 < \frac{R+1}{R+2} - \frac{R}{R+1} < 1$, we get $-\frac{R+1}{R+2} < t_1 \leq t_1 + \tau < R+T+1$. Pick any function φ_1 as in the statement of the lemma. Since $\|x_1\| \leq \|x_0\| + \overline{\eta}_0 < R+1$ we have $\|\varphi_1(t_1 + h)\| < m(R+1)$ $\forall h \geq 0$. Since $\|\varphi_1(t_1 + \tau)\| > \frac{1}{q}$, we must have $\|\varphi_1(t_1 + h)\| > m^{-1}(\frac{1}{q})$ for $0 \leq h \leq \tau$. The remaining part of the lemma is an easy consequence of (4.41)-(4.43). ∎

Let $(t_0, x_0) \in [-\frac{R}{R+1}, R] \times \overline{B}_R$ be fixed. In what follows, we always assume that

$$0 < \eta_0 < \frac{\overline{\eta}_0}{2M + 1} \tag{4.44}$$

We have to distinguish two cases: $V_q(t_0, x_0) \neq 0$ or $V_q(t_0, x_0) = 0$.

First case: $V_q(t_0, x_0) \neq 0$.

We first prove that $V_q(t, x) \neq 0$ in a neighborhood of (t_0, x_0).

CLAIM 8. *If η_0 is small enough, then $V_q(t, x) \neq 0$ for each pair $(t, x) \in \overline{B}_{\eta_0}(t_0, x_0)$.*

Proof. Let $\varphi_0 \in \mathcal{S}_{t_0, x_0}$ be such that $V_q(t_0, x_0) - e^{2\lambda\tau}G_q(\|\varphi_0(t_0 + \tau)\|) < \frac{V_q(t_0, x_0)}{2}$ for some $\tau \in (0, T]$. Hence $\|\varphi_0(t_0 + \tau)\| > \frac{1}{q}$ and it follows from Lemma 4.8 that

$$m^{-1}\left(\frac{1}{q}\right) < \|\varphi_0(t_0 + h)\| < m(R+1) \quad \text{for } 0 \leq h \leq \tau. \tag{4.45}$$

By decreasing η_0 if needed, we may assume that $\eta_0 < \tau$, that φ_0 is defined on $[t_0 - \eta_0, t_0]$, that (4.45) holds true for $-\eta_0 \leq h \leq \tau$ and that $[t_0 - \eta_0, t_0 + \tau] \subset [-\frac{R+1}{R+2}, R+T+1]$. Let $(t, x) \in \overline{B}_{\eta_0}(t_0, x_0)$. Then $|t - t_0| \leq \eta_0$ (hence $\varphi_0(t)$ is meaningful) and, by (4.39),

$$\begin{aligned}\|\varphi_0(t) - x\| &\leq \|\varphi_0(t) - \varphi_0(t_0)\| + \|x_0 - x\| \\ &\leq M|t - t_0| + \eta_0 \\ &\leq (M+1)\eta_0 < \overline{\eta}_0 < b,\end{aligned}$$

by (4.44). By Lemmas 4.1 and 4.8, there exists $\psi \in \mathcal{S}_{t,x}$ such that

$$\|\varphi_0(t+s) - \psi(t+s)\| \leq \|\varphi_0(t) - x\| e^{K|s|} \quad (4.46)$$

as long as $m^{-1}(\frac{1}{q}) \leq \|\varphi_0(t+s)\| \leq m(R+1)$ and the right hand side of (4.46) is less than b. This occurs for any s such that $t_0 \leq t+s \leq t_0 + \tau$, since

$$\|\varphi_0(t) - x\| e^{K|s|} \leq \bar{\eta}_0 e^{K(T+1)} = b.$$

We infer from (4.46) that

$$\begin{aligned}\|\psi(t_0+\tau)\| &\geq \|\varphi_0(t_0+\tau)\| - \|\varphi_0(t) - x\| e^{K(\eta_0+\tau)} \\ &\geq \|\varphi_0(t_0+\tau)\| - (M+1) e^{K(T+1)} \eta_0 \\ &> \tfrac{1}{q},\end{aligned}$$

if η_0 is small enough. This implies $V_q(t,x) > 0$, as required. ∎

Let η_0 be as in Claim 8, and let $(t_1,x_1),(t_2,x_2) \in \overline{B}_{\eta_0}(t_0,x_0)$. (4.37) is a particular instance of next

CLAIM 9. $\quad |V_q(t_1,x_1) - V_q(t_2,x_2)| \leq L \|(t_1-t_2, x_1-x_2)\|.$

Without loss of generality we may assume that $t_1 \leq t_2$. We need some preliminary steps.

<u>Step 1.</u> $\quad |V_q(t_1,x_1) - V_q(t_1,x_2)| \leq e^{(2\lambda+K)T} \|x_1-x_2\|.$

Proof. By definition of $V_q(t_1,x_1)$, for each $\sigma \in (0, V_q(t_1,x_1))$, there exists a solution $\varphi_1 \in \mathcal{S}_{t_1,x_1}$ and $\tau \in [0,T]$ s.t.

$$V_q(t_1,x_1) - \sigma < e^{2\lambda\tau} G_q(\|\varphi_1(t_1+\tau)\|) \leq V_q(t_1,x_1).$$

Hence $V_q(t_1,x_1) - V_q(t_1,x_2) < e^{2\lambda\tau} G_q(\|\varphi_1(t_1+\tau)\|) - V_q(t_1,x_2) + \sigma$. Since $\|x_1-x_2\| \leq 2\eta_0 < \bar{\eta}_0 < b$, we infer from Lemmas 4.1, 4.8 and Claim 8 that there exists a solution $\varphi_2 \in \mathcal{S}_{t_1,x_2}$ such that $\|\varphi_1(t) - \varphi_2(t)\| \leq \|x_1-x_2\| e^{K|t-t_1|}$, as long as $m^{-1}(\frac{1}{q}) \leq \|\varphi_1(t)\| \leq m(R+1)$ and $\|x_1-x_2\| e^{K|t-t_1|} \leq b$, hence for $t_1 \leq t \leq t_1+\tau$. Since $V_q(t_1,x_2) \geq e^{2\lambda\tau} G_q(\|\varphi_2(t_1+\tau)\|)$ and since G_q is Lipschitz continuous with constant 1, we get

$$\begin{aligned}V_q(t_1,x_1) - V_q(t_1,x_2) &\leq e^{2\lambda\tau} \Big(G_q(\|\varphi_1(t_1+\tau)\|) - G_q(\|\varphi_2(t_1+\tau)\|)\Big) + \sigma \\ &\leq e^{(2\lambda+K)T} \|x_1-x_2\| + \sigma.\end{aligned}$$

140   Differential inclusions

The reasoning can be repeated by exchanging the roles of x_1 and x_2. σ being arbitrary, the proof of Step 1 is completed. ∎

Step 2. $V_q(t_2, x_2) - V_q(t_1, x_2) \leq M e^{(2\lambda+K)T} |t_2 - t_1|.$

Proof. Pick any $\varphi \in \mathcal{S}_{t_1,x_2}$. Set $x_3 = \varphi(t_2)$. Due to Lemma 4.6,

$$V_q(t_2, x_3) \leq e^{-2\lambda(t_2-t_1)} V_q(t_1, x_2) \leq V_q(t_1, x_2),$$

hence

$$V_q(t_2, x_2) - V_q(t_1, x_2) \leq V_q(t_2, x_2) - V_q(t_2, x_3). \tag{4.47}$$

Notice that

$$\|x_2 - x_3\| \leq \int_{t_1}^{t_2} \|\dot{\varphi}(t)\| \, dt \leq M|t_2 - t_1| \leq 2\eta_0 M.$$

Since, by (4.44), $2\eta_0 M < \bar{\eta}_0 < b$ we may use the same arguments as in Step 1 to conclude that

$$V_q(t_2, x_2) - V_q(t_2, x_3) \leq e^{(2\lambda+K)T} \|x_2 - x_3\| \leq M e^{(2\lambda+K)T} |t_2 - t_1|$$

which, combined with (4.47), gives the desired result. ∎

Step 3. $V_q(t_1, x_2) - V_q(t_2, x_2) \leq \left(M e^{KT} + 2\lambda m(R+1) \right) e^{2\lambda T} |t_1 - t_2|.$

Proof. By definition, for each $\sigma \in (0, V_q(t_1, x_2))$, there exists a solution $\psi \in \mathcal{S}_{t_1,x_2}$ and $\tau \in [0,T]$ such that $V_q(t_1, x_2) \leq e^{2\lambda\tau} G_q(\|\psi(t_1+\tau)\|) + \sigma$. Let us distinguish two cases.

First case: $t_1 + \tau > t_2 \, (\geq t_1)$.
Set $x_4 := \psi(t_2)$. Then $\|x_4 - x_2\| \leq \int_{t_1}^{t_2} \|\dot{\psi}(t)\| \, dt \leq M|t_2 - t_1| < 2M\eta_0$, hence $\|x_4 - x_0\| < (2M+1)\eta_0 < \bar{\eta}_0$ (by (4.44)), and we get $(t_2, x_4) \in \overline{B}_{\bar{\eta}_0}(t_0, x_0)$. It follows from Lemma 4.8 (with (t_2, x_4) substituted for (t_1, x_1)) combined with Lemma 4.1 that there exists a solution $\varphi \in \mathcal{S}_{t_2, x_2}$ such that

$$\|\varphi(t) - \psi(t)\| \leq \|x_4 - x_2\| e^{K|t-t_2|} \qquad \text{for } t_2 \leq t \leq t_1 + \tau.$$

Since $V_q(t_2, x_2) \geq e^{2\lambda(t_1+\tau-t_2)} G_q(\|\varphi(t_1+\tau)\|)$, we may write

$$\begin{aligned}V_q(t_1, x_2) - V_q(t_2, x_2) &\leq e^{2\lambda\tau} G_q(\|\psi(t_1+\tau)\|) - e^{2\lambda(\tau+t_1-t_2)} G_q(\|\varphi(t_1+\tau)\|) + \sigma \\ &\leq e^{2\lambda\tau} \{|G_q(\|\psi(t_1+\tau)\|) - G_q(\|\varphi(t_1+\tau)\|)| \\ &\quad + (1 - e^{-2\lambda|t_1-t_2|}) G_q(\|\varphi(t_1+\tau)\|)\} + \sigma.\end{aligned}$$

But

$$\begin{aligned}\left| G_q(\|\psi(t_1+\tau)\|) - G_q(\|\varphi(t_1+\tau)\|) \right| &\leq \|\psi(t_1+\tau) - \varphi(t_1+\tau)\| \\ &\leq \|x_4 - x_2\| e^{K|t_1+\tau-t_2|} \\ &\leq Me^{KT} |t_2 - t_1|\end{aligned}$$

and $(1 - e^{-2\lambda|t_1-t_2|}) G_q(\|\varphi(t_1+\tau)\|) \leq 2\lambda|t_1 - t_2| m(R+1)$. Summing up we get

$$V_q(t_1, x_2) - V_q(t_2, x_2) \leq e^{2\lambda T}(Me^{KT} + 2\lambda m(R+1))|t_1 - t_2| + \sigma. \quad (4.48)$$

Second case: $t_1 + \tau \leq t_2$.
Using $V_q(t_2, x_2) \geq G_q(\|x_2\|)$ we get

$$\begin{aligned}V_q(t_1, x_2) - V_q(t_2, x_2) &\leq e^{2\lambda\tau} G_q(\|\psi(t_1+\tau)\|) - G_q(\|x_2\|) + \sigma \\ &\leq e^{2\lambda\tau} |G_q(\|\psi(t_1+\tau)\|) - G_q(\|x_2\|)| \\ &\quad + (e^{2\lambda\tau} - 1) G_q(\|x_2\|) + \sigma.\end{aligned}$$

Since $|\|\psi(t_1+\tau)\| - \|x_2\|| \leq \|\int_{t_1}^{t_1+\tau} \dot\psi(t)\,dt\| \leq M\tau \leq M|t_2-t_1|$ and $|e^{2\lambda\tau} - 1| \leq e^{2\lambda\tau} \cdot 2\lambda\tau \leq 2\lambda e^{2\lambda T}|t_2 - t_1|$, we finally obtain

$$\begin{aligned}V_q(t_1, x_2) - V_q(t_2, x_2) &\leq e^{2\lambda T}(M + 2\lambda m(R+1))|t_1 - t_2| + \sigma \\ &\leq e^{2\lambda T}(Me^{KT} + 2\lambda m(R+1))|t_1 - t_2| + \sigma.\end{aligned}$$

Hence (4.48) holds in both cases and the proof of Step 3 is completed in letting $\sigma \to 0$. Finally, by Steps 1-3,

$$\begin{aligned}|V_q(t_1, x_1) - V_q(t_2, x_2)| &\leq |V_q(t_1, x_1) - V_q(t_1, x_2)| + |V_q(t_1, x_2) - V_q(t_2, x_2)| \\ &\leq e^{(2\lambda+K)T} \|x_1 - x_2\| \\ &\quad + (Me^{KT} + 2\lambda m(R+1))e^{2\lambda T}|t_1 - t_2| \\ &\leq L\|(t_1 - t_2, x_1 - x_2)\|.\end{aligned}$$

This finishes the proof of Claim 9 and the one of Proposition 4.6 when $V_q(t_0, x_0) \neq 0$. We now proceed to the

Second case: $V_q(t_0, x_0) = 0$

It follows from (4.44) that $M\eta_0 < 1$. We claim that for any $(t, x) \in \overline{B}_{\eta_0}(t_0, x_0)$ and any $\varphi \in \mathcal{S}_{t,x}$, φ is defined on $[t - \eta_0, +\infty)$ (hence at time t_0). Indeed, by (4.39), for any $s \in \text{dom}(\varphi) \cap [t - \eta_0, t]$

$$\|\varphi(s)\| \leq \|\varphi(s) - \varphi(t)\| + \|x\| \leq M|s-t| + R + \eta_0 \leq (M+1)\eta_0 + R \leq R+1, \quad (4.49)$$

as long as $\|\varphi(s)\| \leq m(R+2)$ (a condition which must be satisfied in order to apply (4.39)). (Notice that $[t - \eta_0, t] \subset [-\frac{R+1}{R+2}, R+1]$, since $t - \eta_0 \geq t_0 - 2\eta_0 > t_0 - \overline{\eta}_0 > -\frac{R+1}{R+2}$.) Since $R + 1 < m(R+2)$, a standard argument shows that $[t - \eta_0, t] \subset \text{dom}(\varphi)$ and that (4.49) holds true on $[t - \eta_0, t]$. Pick any $(t, x) \in \overline{B}_{\eta_0}(t_0, x_0)$. If $V_q(t, x) = 0$, then (4.37) is trivial. If $V_q(t, x) > 0$, then for any $\sigma \in (0, V_q(t, x))$ we may find a solution $\varphi \in \mathcal{S}_{t,x}$ and a time $\tau \in [0, T]$ such that

$$V_q(t, x) \leq e^{2\lambda \tau} G_q(\|\varphi(t+\tau)\|) + \sigma. \quad (4.50)$$

We again distinguish two cases.

First case: $t_0 < t + \tau$

Recall that φ is defined on $[t - \eta_0, +\infty)$, hence at time t_0. Since $\|\varphi(t+\tau)\| > \frac{1}{q}$, we still have $m^{-1}(\frac{1}{q}) < \|\varphi(s)\| < m(R+1)$ for any $s \in [t - \eta_0, t + \tau]$. Furthermore,

$$\|\varphi(t_0) - x_0\| \leq \|\varphi(t_0) - \varphi(t)\| + \|x - x_0\| \leq (M+1)\eta_0 < \overline{\eta}_0,$$

according to (4.44). By Lemmas 4.1 and 4.8 there exists a solution $\psi \in \mathcal{S}_{t_0, x_0}$ such that

$$\begin{aligned}
\|\psi(t+\tau) - \varphi(t+\tau)\| &\leq \|\psi(t_0) - \varphi(t_0)\| e^{K|t+\tau - t_0|} \\
&\leq \big(\|x_0 - x\| + \|\varphi(t) - \varphi(t_0)\|\big) e^{K(T+1)} \\
&\leq (M+1) e^{K(T+1)} \|(t - t_0, x - x_0)\|.
\end{aligned}$$

Notice that $G_q(\|\psi(t+\tau)\|) = 0$, since $V_q(t_0, x_0) = 0$. We obtain

$$\begin{aligned}
V_q(t, x) &\leq e^{2\lambda \tau} \big[G_q(\|\varphi(t+\tau)\|) - G_q(\|\psi(t+\tau)\|)\big] + \sigma \\
&\leq (M+1) e^{2\lambda T + K(T+1)} \|(t - t_0, x - x_0)\| + \sigma.
\end{aligned} \quad (4.51)$$

Second case: $t_0 \geq t + \tau$

Since

$$\begin{aligned}
G_q(\|\varphi(t+\tau)\|) &= G_q(\|\varphi(t+\tau)\|) - G_q(\|x_0\|) \\
&\leq \big|G_q(\|\varphi(t+\tau)\|) - G_q(\|x\|)\big| + \big|G_q(\|x\|) - G_q(\|x_0\|)\big| \\
&\leq M\tau + \|x - x_0\| \leq M|t - t_0| + \|x - x_0\|
\end{aligned} \quad (4.52)$$

we have, by (4.50) and (4.52)

$$V_q(t,x) \le e^{2\lambda T}(M+1)\|(t-t_0, x-x_0)\| + \sigma.$$

Thus, in each case, $0 \le V_q(t,x) \le L\|(t-t_0, x-x_0)\| + \sigma$. Letting $\sigma \to 0$ we get (4.37). The proof of Proposition 4.6 is achieved. This completes Step 2 in the proof of Theorem 4.5.

THIRD STEP. (Regularization of V_L)

The regularization of V_L is carried out by means of convolution with mollifiers and of some partition of unity. Next lemma is similar to [38, Lem. 5.1] and [118, Lem. 1].

Lemma 4.9 *Let S be any compact set in $U = (-1, +\infty) \times (\mathbb{R}^n \setminus \{0\})$ and let $\varepsilon > 0$. Then there exists a function $\overline{V}$ of class C^∞ and with compact support in $(-1, +\infty) \times \mathbb{R}^n$, such that*

$$\|\overline{V} - V_L\|_{L^\infty(S)} < \varepsilon \quad \text{and} \quad (4.53)$$

$$\forall (t_0, x_0) \in S, \; \forall v \in F_2(t_0, x_0),$$
$$\frac{\partial \overline{V}}{\partial t}(t_0, x_0) + \langle \nabla_x \overline{V}(t_0, x_0), v \rangle \le -\frac{3}{2}\lambda V_L(t_0, x_0). \quad (4.54)$$

Proof. Let $\rho \in C^\infty(\mathbb{R}^{n+1}, \mathbb{R})$ be a nonnegative function such that $\operatorname{supp}(\rho) \subset \overline{B}_1$ and $\int_{\mathbb{R}^{n+1}} \rho(t,x)\,dt\,dx = 1$. For any $\delta > 0$, set $\rho_\delta(t,x) := \frac{1}{\delta^{n+1}} \rho\left(\frac{t}{\delta}, \frac{x}{\delta}\right)$ and

$$\begin{aligned} V_\delta(t,x) &:= V_L * \rho_\delta(t,x) \\ &= \int_{\mathbb{R}^{n+1}} V_L(t-s, x-y)\rho_\delta(s,y)\,ds\,dy \\ &= \int_{\|(\bar{s},\bar{y})\|\le 1} V_L(t-\delta \bar{s}, x-\delta \bar{y})\rho(\bar{s},\bar{y})\,d\bar{s}\,d\bar{y}. \end{aligned}$$

Then V_δ is defined and of class C^∞ on $(-1+\delta, +\infty) \times \mathbb{R}^n$, and $V_\delta(t,x) \to V_L(t,x)$ uniformly on S as $\delta \to 0$. If θ denotes a function of class C^∞ with compact support in $(-1, +\infty) \times \mathbb{R}^n$ and which assumes the value 1 on a neighborhood of S, then the (smooth) function $\overline{V} := \theta \cdot V_\delta$ has a compact support in $(-1, +\infty) \times \mathbb{R}^n$ and it satisfies (4.53), provided that δ be small enough. To complete the proof of the lemma, it remains to show that there exists $\delta_0 > 0$ such that for any $\delta \in (0, \delta_0)$,

$$\forall (t_0, x_0) \in S, \; \forall v \in F_2(t_0, x_0)$$
$$\frac{\partial V_\delta}{\partial t}(t_0, x_0) + \langle \nabla_x V_\delta(t_0, x_0), v \rangle \le -\frac{3}{2}\lambda V_L(t_0, x_0).$$

Let $\delta_1 > 0$ be such that $S + \overline{B}_{\delta_1} \subset U$ and let $L > 0$ be such that for any pairs $(t_1, x_1), (t_2, x_2)$ in $S + \overline{B}_{\delta_1}$ (h denoting the Hausdorff distance between closed sets)

$$h\big(F_2(t_1, x_1), F_2(t_2, x_2)\big) + |V_L(t_1, x_1) - V_L(t_2, x_2)| \leq L \, \|(t_1 - t_2, x_1 - x_2)\|. \tag{4.55}$$

It follows that for a.e. $(t, x) \in S + B_{\delta_1}$, V_L is differentiable at (t, x) and

$$\left\| \left(\frac{\partial V_L}{\partial t}, \nabla_x V_L \right)(t, x) \right\| \leq L. \tag{4.56}$$

Let $\delta \in (0, \delta_1)$, $(t_0, x_0) \in S$ and $v \in F_2(t_0, x_0)$. Applying Lebesgue's theorem we infer from (4.55) that

$$\begin{aligned}
&\tfrac{\partial V_\delta}{\partial t}(t_0, x_0) + \langle \nabla_x V_\delta(t_0, x_0), v \rangle \\
&= \lim_{\eta \to 0} \int_{\|(\bar{s}, \bar{y})\| \leq 1} \tfrac{1}{\eta} \big[V_L(t_0 - \delta \bar{s} + \eta, x_0 - \delta \bar{y} + \eta v) \\
&\qquad\qquad - V_L(t_0 - \delta \bar{s}, x_0 - \delta \bar{y}) \big] \rho(\bar{s}, \bar{y}) \, d\bar{s} \, d\bar{y} \\
&= \int_{\|(\bar{s}, \bar{y})\| \leq 1} \Big(\tfrac{\partial V_L}{\partial t}(t_0 - \delta \bar{s}, x_0 - \delta \bar{y}) \\
&\qquad\qquad + \langle \nabla_x V_L(t_0 - \delta \bar{s}, x_0 - \delta \bar{y}), v \rangle \Big) \rho(\bar{s}, \bar{y}) \, d\bar{s} \, d\bar{y}.
\end{aligned} \tag{4.57}$$

g denoting also the map defined by (4.35), we may write, thanks to (4.35), (4.56), (4.57) and Corollary 4.3,

$$\begin{aligned}
&\tfrac{\partial V_\delta}{\partial t}(t_0, x_0) + \langle \nabla_x V_\delta(t_0, x_0), v \rangle \\
&= \int_{\|(\bar{s}, \bar{y})\| \leq 1} \left(\tfrac{\partial V_L}{\partial t} + \langle \nabla_x V_L, g \rangle \right)(t_0 - \delta \bar{s}, x_0 - \delta \bar{y}) \rho(\bar{s}, \bar{y}) \, d\bar{s} \, d\bar{y} \\
&\quad + \int_{\|(\bar{s}, \bar{y})\| \leq 1} \langle \nabla_x V_L(t_0 - \delta \bar{s}, x_0 - \delta \bar{y}), v - g(t_0 - \delta \bar{s}, x_0 - \delta \bar{y}) \rangle \rho(\bar{s}, \bar{y}) \, d\bar{s} \, d\bar{y}, \\
&\leq -2\lambda \, V_\delta(t_0, x_0) + \sqrt{n} \, L \int_{\|(\bar{s}, \bar{y})\| \leq 1} \|v - g(t_0 - \delta \bar{s}, x_0 - \delta \bar{y})\| \rho(\bar{s}, \bar{y}) \, d\bar{s} \, d\bar{y}.
\end{aligned} \tag{4.58}$$

Now, observe that for any $\|(\bar{s}, \bar{y})\| \leq 1$,

$$\begin{aligned}
\|v - g(t_0 - \delta \bar{s}, x_0 - \delta \bar{y})\| &= \mathrm{dist}\big(v, F_2(t_0 - \delta \bar{s}, x_0 - \delta \bar{y})\big) \\
&\leq h\big(F_2(t_0, x_0), F_2(t_0 - \delta \bar{s}, x_0 - \delta \bar{y})\big) \\
&\leq L \delta,
\end{aligned}$$

due to (4.55). We obtain

$$\frac{\partial V_\delta}{\partial t}(t_0, x_0) + \langle \nabla_x V_\delta(t_0, x_0), v \rangle \leq -2\lambda \, V_\delta(t_0, x_0) + \sqrt{n} \, L^2 \, \delta \leq -\frac{3}{2} \lambda V_L(t_0, x_0),$$

if δ is small enough. ∎

Let $(\psi_i)_{i \geq 1}$ be a C^∞ partition of unity on U, such that for each $i \geq 1$ the support S_i of ψ_i is a *compact* set in U. For each $i \geq 1$, set

$$\begin{aligned}
q_i &:= \sup_{(t,x) \in S_i, \, v \in F_2(t,x)} \left| \tfrac{\partial \psi_i}{\partial t}(t, x) + \langle \nabla_x \psi_i(t, x), v \rangle \right| < +\infty \\
\varepsilon_i &:= \tfrac{1}{2^{i+2}(1 + q_i)} \cdot \tfrac{\lambda}{\lambda + 1} \cdot \min_{(t,x) \in S_i} V_L(t, x) > 0.
\end{aligned}$$

Lemma 4.9 provides, for each $i \geq 1$, a function V_i of class C^∞ on $(-1, +\infty) \times \mathbb{R}^n$ such that
$$\forall (t,x) \in S_i, \quad |V_i(t,x) - V_L(t,x)| < \varepsilon_i$$
and
$$\forall (t,x) \in S_i, \ \forall v \in F_2(t,x), \quad \frac{\partial V_i}{\partial t}(t,x) + \langle \nabla_x V_i(t,x), v \rangle \leq -\frac{3}{2} \lambda V_L(t,x).$$

We define on $(-1, +\infty) \times \mathbb{R}^n$ a function $\tilde{V}$ by
$$\tilde{V}(t,x) := \begin{cases} \sum_{i=1}^{+\infty} \psi_i(t,x) V_i(t,x) & \text{if } x \neq 0, \\ 0 & \text{if } x = 0. \end{cases}$$

It is clear that $\tilde{V}$ is of class C^∞ on U. Moreover, for each $(t,x) \in U$,
$$\begin{aligned} |\tilde{V}(t,x) - V_L(t,x)| &\leq \sum_{i=1}^{+\infty} \psi_i(t,x) |V_i(t,x) - V_L(t,x)| \\ &\leq \sum_{i=1}^{+\infty} \frac{1}{2^{i+2}} V_L(t,x) = \frac{1}{4} V_L(t,x), \end{aligned}$$
hence
$$\frac{3}{4} a_L(\|x\|) \leq \frac{3}{4} V_L(t,x) \leq \tilde{V}(t,x) \leq \frac{5}{4} V_L(t,x) \leq \frac{5}{4} b_L(\|x\|),$$
where $a_L, b_L \in \mathcal{K}_0^\infty$ are defined in Claims 4 and 6, respectively. Clearly, $\tilde{V}$ is continuous on $(-1, +\infty) \times \mathbb{R}^n$. Pick any $(t,x) \in U$ and any $v \in F_2(t,x)$. Then
$$\begin{aligned} \frac{\partial \tilde{V}}{\partial t}(t,x) + \langle \nabla_x \tilde{V}(t,x), v \rangle &= \sum_{i=1}^{+\infty} \left(\frac{\partial \psi_i}{\partial t} + \langle \nabla_x \psi_i, v \rangle \right)(V_i - V_L) \\ &\quad + \sum_{i=1}^{+\infty} \psi_i \left(\frac{\partial V_i}{\partial t} + \langle \nabla_x V_i, v \rangle \right) \\ &\leq \sum_{i=1}^{+\infty} \left[\frac{q_i}{1+q_i} \cdot \frac{\lambda}{2^{i+2}} \cdot V_L(t,x) \right] \\ &\quad + \sum_{i=1}^{+\infty} \psi_i \left(-\frac{3}{2} \lambda V_L(t,x) \right) \\ &\leq -\frac{5}{4} \lambda V_L(t,x) \leq -\lambda \tilde{V}(t,x). \end{aligned}$$

In order to get a function which is smooth *up to* $x = 0$, we apply [91, Thm. 6], which asserts the existence of a function $\nu \in C^\infty(\mathbb{R})$ such that $\nu(r) = 0$ for $r \leq 0$, $\nu'|_{\mathbb{R}^+}$ is increasing (hence $\nu'(r) > 0$, $\nu(r) > 0$ for $r > 0$), $\nu(r) \to +\infty$ as $r \to +\infty$ and $\nu \circ \tilde{V} \in C^\infty((-1, +\infty) \times \mathbb{R}^n)$. Moreover, $\partial^\alpha(\nu \circ \tilde{V})(t_0, 0) = 0$ for any $t_0 > -1$ and any $\alpha \in \mathbb{N}^{n+1}$. We set $V := \nu \circ \tilde{V}$ and $a(r) := \nu\left(\frac{3}{4} a_L(r)\right)$, $b(r) := \nu\left(\frac{5}{4} b_L(r)\right)$ for all $r > 0$. Clearly, $a, b \in \mathcal{K}_0^\infty$ and (4.7) holds true. Let $(t,x) \in ([0, +\infty) \setminus N_0) \times (\mathbb{R}^n \setminus \{0\})$ and let $v \in F(t,x)$. According to Claim 2, $v \in F_2(t,x)$, hence
$$\frac{\partial \tilde{V}}{\partial t}(t,x) + \langle \nabla_x \tilde{V}(t,x), v \rangle \leq -\lambda \tilde{V}(t,x).$$

On the other hand, since ν' is increasing on $\mathbb{R}^+$, we get for each $r \geq 0$
$$\nu(r) = \int_0^r \nu'(s)\, ds \leq \int_0^r \nu'(r)\, ds = r\, \nu'(r).$$

It follows that

$$\frac{\partial V}{\partial t}(t,x) + \langle \nabla_x V(t,x), v\rangle = \nu'(\tilde{V}(t,x)) \cdot \left(\frac{\partial \tilde{V}}{\partial t}(t,x) + \langle \nabla_x \tilde{V}(t,x), v\rangle\right)$$
$$\leq -\lambda \nu'(\tilde{V}(t,x))\tilde{V}(t,x)$$
$$\leq -\lambda \nu(\tilde{V}(t,x)) = -\lambda V(t,x).$$

The proof of Theorem 4.5 is complete in the general setting. ■

Let us briefly indicate the modifications to be brought in the time periodic or time independent cases. We begin with the

- TIME INDEPENDENT CASE.

Lemma 4.3 holds true when we replace (4.13) by $\dot{x} \in \overline{\mathrm{co}}\{F(\overline{B}_\delta(x))\}$ and the condition $t_0 \in [k, k+1]$ by $t_0 = 0$. Here, δ depends only on i, so we write δ_i instead of δ_i^k. Next, we pick a Lipschitz continuous function (with 1 as a Lipschitz constant) $\delta : \mathbb{R}^n \to \mathbb{R}^+$, such that $\delta(x) = 0 \iff x = 0$ and $\delta(x) < \min(\delta_i, \delta_{p_i})$ whenever $2^{p_i} \leq \|x\| \leq \varphi_i(0)$. F_1 is defined by $F_1(x) := \overline{\mathrm{co}}\{F(\overline{B}_{\delta(x)}(x))\}$. We set $U := \mathbb{R}^n \setminus \{0\}$, and, for any $x \in U$, $W(x) := B_{\frac{1}{3}\delta(x)}(x)$. We pick a C^∞ partition of unity $(\rho_i)_{i \geq 1}$ subordinate to the open covering $(W(x))_{x \in U}$ of U. For each $i \geq 1$, we choose some $x_i \neq 0$ such that $\mathrm{supp}(\rho_i) \subset W(x_i)$. F_2 is defined by

$$F_2(x) := \begin{cases} \sum_{i=1}^\infty \rho_i(x) \, \overline{\mathrm{co}}\{F(\overline{B}_{\frac{1}{3}\delta(x_i)}(x_i))\} & \text{if } x \neq 0, \\ F(0) & \text{if } x = 0. \end{cases}$$

F_2 is time independent. The same is true for the functions V_q and V_L, defined by (4.23), (4.26), respectively. Lemma 4.9 remains valid when S is a compact set in $\mathbb{R}^n \setminus \{0\}$ and (4.54) is replaced by $\langle \nabla_x \overline{V}(x_0), v\rangle \leq -\frac{3}{2}\lambda V_L(x_0)$ for all $x_0 \in S$ and all $v \in F_2(x_0)$. Here, the function V provided by Lemma 4.9 is of class C^∞ on $\mathbb{R}^n$. We set $S_i := \mathrm{supp}(\rho_i)$, and q_i, ε_i are defined as in the general case, with $\rho_i(x)$ instead of $\psi_i(t,x)$. Finally, we set

$$\tilde{V}(x) := \begin{cases} \sum_{i=1}^{+\infty} \rho_i(x) V_i(x) & \text{if } x \neq 0, \\ 0 & \text{if } x = 0, \end{cases}$$

and we define again V by $V := \nu \circ \tilde{V}$. We now turn to the

- TIME PERIODIC CASE.

Without loss of generality, we may assume that the period is $T = 1$, that F is defined on the whole space $\mathbb{R}_t \times \mathbb{R}_x^n$ and that $(\mathbf{H}')$ holds true (on $\mathbb{R}^{n+1}$) for some $N_0 \subset \mathbb{R}$ of measure 0 and such that $N_0 + 1 = N_0$. Here, we take

$E := (\mathbb{R} \setminus N_0) \times \mathbb{R}^n$ and $U := \mathbb{R} \times (\mathbb{R}^n \setminus \{0\})$. If $x(\cdot)$ is any solution of (4.1) issuing from x_0 at some time $t_0 \in \mathbb{R}$, then $\tilde{x}(t) := x(t + [t_0])$ is also a solution of (4.1), issuing from x_0 at time $\tilde{t}_0 := t_0 - [t_0] \in [0, 1]$. Therefore, we may always assume that $t_0 \in [0, 1]$. Thanks to this trivial remark, we see that Lemma 4.3 is valid with a constant $\delta > 0$ which depends *only* on i (and is denoted by δ_i), whatever be $t_0 \in \mathbb{R}$. Next, the function $\delta(x)$ is chosen as in the time-independent case. F_1 is defined by

$$F_1(t, x) := \overline{\text{co}}\{F(\overline{B}_{\delta(x)}(t, x) \cap E)\}.$$

Clearly, F_1 is still 1-periodic in time. In order to construct a multivalued map F_2 which is also 1-periodic in time, we need some partition of unity which is left "invariant" by the transformation $t \mapsto t + 1$. Let $(x_i)_{i \geq 1}$ and $(\rho_i)_{i \geq 1}$ be as in the time-independent case. Pick a function $\omega \in C^\infty([0, 1], [0, 1])$ such that

$$\omega(t) := \begin{cases} 1 & \text{for } 0 \leq t \leq \frac{1}{3}, \\ 0 & \text{for } \frac{2}{3} \leq t \leq 1. \end{cases}$$

Extend ω to $\mathbb{R}$ by setting $\omega(-t) := 1 - \omega(1 - t)$ for $0 \leq t \leq 1$, and $\omega(t) = 0$ for $|t| > 1$. Obviously, $\omega \in \mathcal{D}(\mathbb{R})$ and $(\omega(\cdot - k))_{k \in \mathbb{Z}}$ constitutes a C^∞ partition of unity subordinate to the covering $((k-1, k+1))_{k \in \mathbb{Z}}$ of $\mathbb{R}$. For each $i \geq 1$, pick an integer $M_i > \frac{6}{\delta(x_i)}$. For each $k = 0, \ldots, M_i - 1$, pick a time $t_{i,k} \in \left(\frac{k-1}{M_i}, \frac{k+1}{M_i}\right) \setminus N_0$, and set $t_{i,k+pM_i} := t_{i,k} + p$ for any $0 \leq k \leq M_i - 1$ and any $p \in \mathbb{Z}$. Notice that $(t_{i,k}, x_i) \in U \cap E$ for each pair (i, k). Set $\psi_{i,k}(t, x) := \omega(M_i t - k) \rho_i(x)$ for all $(t, x) \in U$. Then $(\psi_{i,k})_{i \geq 1, k \in \mathbb{Z}}$ is a C^∞ partition of unity subordinate to the covering $\left(\left(\frac{k-1}{M_i}, \frac{k+1}{M_i}\right) \times W(x_i)\right)_{i \geq 1, k \in \mathbb{Z}}$ of U. We are now ready to define F_2.

$$F_2(t, x) := \begin{cases} \sum_{i=1}^\infty \sum_{k \in \mathbb{Z}} \psi_{i,k}(t, x) \, \overline{\text{co}}\{F(\overline{B}_{\frac{1}{3}\delta(x_i)}(t_{i,k}, x_i) \cap E)\} & \text{if } x \neq 0, \\ F(t, 0) & \text{if } x = 0. \end{cases}$$

Let us check that Claim 2 remains valid. Let $x \neq 0$ and let $t \in \mathbb{R} \setminus N_0$. Pick any pair (i, k) such that $\psi_{i,k}(t, x) > 0$, hence $(t, x) \in \left(\frac{k-1}{M_i}, \frac{k+1}{M_i}\right) \times B_{\frac{1}{6}\delta(x_i)}(x_i)$. Since $t_{i,k} \in \left(\frac{k-1}{M_i}, \frac{k+1}{M_i}\right)$ and $\frac{1}{M_i} < \frac{1}{6}\delta(x_i)$, we conclude that $\|(t - t_{i,k}, x - x_i)\| < \frac{1}{3}\delta(x_i)$, hence $F(t, x) \subset F(\overline{B}_{\frac{1}{3}\delta(x_i)}(t_{i,k}, x_i) \cap E)$, which implies $F(t, x) \subset F_2(t, x)$. The proof of Claim 3 is as in the general case. Finally, let us check that F_2 is 1-periodic in time. Let $x \neq 0$ and $t \in \mathbb{R}$. Since $\psi_{i,k}(t + 1, x) = \omega(M_i t + M_i - k) \rho_i(x) = \psi_{i, k - M_i}(t, x)$, setting $k' = k - M_i$, we get

$$F_2(t + 1, x) = \sum_{i=1}^{+\infty} \sum_{k' \in \mathbb{Z}} \psi_{i,k'}(t, x) \, \overline{\text{co}}\{F(\overline{B}_{\frac{1}{3}\delta(x_i)}(t_{i,k'+M_i}, x_i) \cap E)\}.$$

But $t_{i,k'+M_i} = t_{i,k'} + 1$, hence $F\bigl(\overline{B}_{\frac{1}{3}\delta(x_i)}(t_{i,k'+M_i}, x_i)\bigr) \cap E) = F\bigl(\overline{B}_{\frac{1}{3}\delta(x_i)}(t_{i,k'}, x_i)\bigr) \cap E)$. It follows that $F_2(t+1, x) = F_2(t, x)$. We readily infer that the functions V_q (for each $q \geq 1$) and V_L, defined on $\mathbb{R} \times \mathbb{R}^n$ by (4.23) and (4.26), respectively, are 1-periodic in time, as well. In the proof of Lemma 4.9, we observe that $V_\delta(t+1, x) = V_\delta(t, x)$ for each $(t, x) \in \mathbb{R}^{n+1}$. This time, we set $\overline{V} := V_\delta$ (for $\delta > 0$ small enough) instead of $\overline{V} := \theta \cdot V_\delta$, so that $\overline{V}$ is 1-periodic in time. For each pair (i, k), let $S_{i,k}$ denote the support of $\psi_{i,k}$ and let $V_{i,k}$ be the (periodic) function (associated with $S_{i,k}$ and $\varepsilon_{i,k}$), which is provided by Lemma 4.9. Thanks to the periodicity of V_L and $V_{i,k}$, we may in fact require that $V_{i,k+pM_i} \equiv V_{i,k}$ for all $p \in \mathbb{Z}$. Setting $\tilde{V} := \sum_{i \geq 1} \sum_{k \in \mathbb{Z}} \psi_{i,k} V_{i,k}$ and $V := \nu \circ \tilde{V}$, we observe that $\tilde{V}$ and V are 1-periodic in time. ∎

4.2 Robust stability

In this section, we are concerned with the first Liapunov theorem and its converses for differential inclusions. Even if system (4.1) is autonomous, i.e., $F = F(x)$, we do not investigate here the existence of a (continuous or smooth) weak Liapunov function of x alone. (The interested reader is referred to Section 2.2.4 for this issue.) Rather, we focus on the existence of a (possibly smooth) time dependent weak Liapunov function $V(t, x)$. The first Liapunov theorem for differential inclusions is given without proof in subsection 4.2.1. In the following subsection we state several converses of first Liapunov theorem. The one which asserts the existence of a *smooth* Liapunov function is proved in subsection 4.2.3. Some application to control theory is given in the final subsection.

4.2.1 Sufficient conditions

Let $h_0 > 0$ and let $F : [0, +\infty) \times B_{h_0} \to \mathbb{R}^n$ be a set valued map such that the solutions to the Cauchy problem

$$\begin{cases} \dot{x} \in F(t, x), \\ x(t_0) = x_0, \end{cases}$$

exist, at least locally, for each pair $(t_0, x_0) \in [0, +\infty) \times B_{h_0}$. We assume that $0 \in F(t, 0)$ for a.e. $t \geq 0$ and we ask whether the origin is locally (robustly) stable. This is indeed the case if there exists a weak Liapunov function (in the small).

Theorem 4.6 (FIRST LIAPUNOV THEOREM) *Let F be as above. Assume that there exist a number $h < h_0$ and a continuous function $V : [0, +\infty) \times B_h \to \mathbb{R}^+$ which is a weak Liapunov function. That is, we have for some functions $a(\cdot), b(\cdot) \in \mathcal{K}_0^\infty$*

$$a(\|x\|) \leq V(t,x) \leq b(\|x\|) \quad \text{for } t \geq 0, \ 0 \leq \|x\| < h, \tag{4.59}$$

and for any solution $x(\cdot)$ of (4.1) defined on some interval I, it holds

$$t_1, t_2 \in I, \ t_1 < t_2 \Rightarrow V(t_2, x(t_2)) \leq V(t_1, x(t_1)). \tag{4.60}$$

Then the origin is robustly stable for (4.1).

If V turns out to be of class C^1, then to fulfill (4.60) it is sufficient (not necessary) that for almost every $t \geq 0$,

$$\frac{\partial V}{\partial t}(t,x) + \langle \nabla_x V(t,x), v \rangle \leq 0, \quad \forall x \in B_h, \ \forall v \in F(t,x). \tag{4.61}$$

The proof of Theorem 4.6 is exactly the same as the one of Theorem 3.6.

Before passing to the converses of Theorem 4.6 in next section, we provide a condition under which a uniformly stable system is actually robustly stable.

Proposition 4.7 *Let $F : [0, +\infty) \times B_{h_0} \to \mathbb{R}^n$ be a set-valued map which fulfills the assumptions of Filippov's Theorem (Theorem 1.5), i.e., $F(t,x)$ is compact valued, measurable with respect to t and locally Lipschitz continuous with respect to x. If the origin is uniformly stable for (4.1), then it is also robustly stable.*

Proof. We have to prove the existence of a sequence $\{G_i\}_{i=0,1,2,...}$ of open sets in $\mathbb{R}^+ \times \mathbb{R}^n$ and of two sequences of real numbers $\{a_i\}_{i=0,1,2,...}$ and $\{b_i\}_{i=0,1,2,...}$ such that

1) $0 < b_{i+1} < a_i \leq b_i$ for each $i = 0, 1, 2, \ldots$, and $b_i \to 0$ as $i \to +\infty$;

2) $\mathbb{R}^+ \times B_{a_i} \subseteq G_i \subseteq \mathbb{R}^+ \times B_{b_i}$ for each $i = 0, 1, 2, \ldots$;

3) for each $i = 0, 1, 2, \ldots$, for each initial pair $(t_0, x_0) \in G_i$, and for each $\varphi \in \mathcal{S}_{t_0, x_0}$ one has $(t, \varphi(t)) \in G_i$ for each $t \geq t_0$.

Let us start the construction with $b_0 < h_0$. Set $\varepsilon = b_0$. According to the uniform stability, we find a number $a_0 := \delta < b_0$. Define G_0 as

$$\{(t,x) : \exists (t_0, x_0) \text{ with } t \geq t_0 \geq 0, x_0 \in B_{a_0}, \exists \varphi \in \mathcal{S}_{t_0, x_0} \text{ s.t. } x = \varphi(t)\}$$

Property 3) is clearly fulfilled. The first inclusion in 2) is trivial and the second one follows by uniform stability. It remains to prove that G_0 is open. Let

$(\bar{t}, \bar{x}) \in G_0$. Let t_0, x_0, φ be such that $t_0 < \bar{t}$, $x_0 \in B_{a_0}$, $\varphi \in \mathcal{S}_{t_0, x_0}$, $\varphi(\bar{t}) = \bar{x}$. Assume that there is a sequence $(t_k, x_k) \to (\bar{t}, \bar{x})$, but $(t_k, x_k) \notin G_0$. Let σ be such that $B_\sigma(x_0) \subset B_{a_0}$. Let $L \in \mathcal{L}^1(t_0, \bar{t}+1)$ be such that

$$\forall t \in [t_0, \bar{t}+1], \ \forall x', x'' \in \overline{B}_{b_0}, \quad h(F(t, x'), F(t, x'')) \leq L(t) \|x' - x''\|,$$

and let $m := \int_{t_0}^{\bar{t}+1} L(\tau) \, d\tau$. By continuity $\exists \eta > 0$ such that for $t \in (\bar{t} - \eta, \bar{t} + \eta)$ one has

$$|\varphi(t) - \bar{x}| < \frac{\sigma e^{-m}}{2}.$$

Let k be such that

$$|\bar{t} - t_k| < \min\{1, \eta\}$$

and

$$|x_k - \bar{x}| < \frac{\sigma e^{-m}}{2}.$$

Now, it is clear that $|x_k - \varphi(t_k)| < \sigma e^{-m}$. Under the assumption about F, we can apply Theorem 1.5 to find $\psi \in \mathcal{S}_{t_k, x_k}$ such that

$$|\psi(t) - \varphi(t)| < \sigma$$

for $t \in [t_0, t_k]$. As a consequence, we have $|\psi(t_0) - x_0| < \sigma$, hence $\psi(t_0) \in B_{a_0}$ and $(t_k, x_k) \in G_0$, a contradiction.

The construction can be iterated by starting each step with a choice of

$$b_i < \min\{a_{i-1}, h_0/2^i\}.$$

∎

We may also wonder if the robust stability of a system is preserved by convexification of the set-valued map. This is indeed the case, as for the asymptotic stability (see above Proposition 4.2).

Proposition 4.8 *Let $F : [0, +\infty) \times B_{h_0} \to \mathbb{R}^n$ be a set-valued map which fulfills the assumptions of Filippov's Theorem (Theorem 1.5). If the origin is robustly stable for*

$$\dot{x} \in F(t, x), \tag{4.62}$$

then the same is true for

$$\dot{x} \in \overline{\mathrm{co}} \, F(t, x). \tag{4.63}$$

Proof. Let $\{a_i\}$, $\{b_i\}$ and $\{G_i\}$ be sequences of real numbers such that
- $0 < b_{i+1} < a_i \leq b_i < h_0$ for each $i = 0, 1, 2, \ldots$, and $b_i \to 0$ as $i \to +\infty$;
- $\mathbb{R}^+ \times B_{a_i} \subseteq G_i \subseteq \mathbb{R}^+ \times B_{b_i}$ for each $i = 0, 1, 2, \ldots$;
- for each $i = 0, 1, 2, \ldots$, for each initial pair $(t_0, x_0) \in G_i$, and for each $\varphi \in S_{t_0, x_0}$ one has $(t, \varphi(t)) \in G_i$ for each $t \geq t_0$.

We claim that the same sequences $\{a_i\}$, $\{b_i\}$, $\{G_i\}$ can be used to check the robust stability of (4.63). Indeed, assume by contradiction that for some $i \geq 1$, some $(t_0, x_0) \in G_i$, some solution $\psi(t)$ of (4.63) and some $T > t_0$ we may have

$$\psi(t_0) = x_0 \quad \text{and} \quad (T, \psi(T)) \in \partial G_i.$$

Let $r_0 > 0$ be such that $\{t_0\} \times B_{r_0}(x_0) \subset G_i$ and $r_0/2 < b_{i-1} - b_i$. Since F is locally Lipschitz continuous with respect to x, there exists a function $L \in \mathcal{L}^1(t_0, T+1)$ such that

$$\forall t \in [t_0, T+1], \forall x', x'' \in \overline{B}_{b_{i-1}}, \quad h(F(t, x'), F(t, x'')) \leq L(t) \|x' - x''\|.$$

Let finally

$$r_1 := \frac{r_0}{8} e^{-\int_{t_0}^{T+1} L(\tau)\, d\tau}.$$

By Theorem [59, Thm. 3], there exists a solution $\tilde{\varphi}(t)$ of (4.62) such that

$$|\psi(t) - \tilde{\varphi}(t)| < r_1 \quad \text{for } t \in [t_0, T].$$

In particular, we have (recall that $\psi(t_0) = x_0$)

$$|x_0 - \tilde{\varphi}(t_0)| < r_1 < \frac{r_0}{2} \quad \text{and} \quad |\psi(T) - \tilde{\varphi}(T)| < r_1,$$

hence $(t_0, \tilde{\varphi}(t_0)) \in G_i$ and $\|\tilde{\varphi}(t)\| < b_i$ for $t \geq t_0$. The solution $\psi(t)$ can be continued on the right of T, and since it is continuous, we can find $\theta \in (0, 1)$ such that

$$|\psi(t) - \tilde{\varphi}(t)| < 2r_1 \quad \text{for } t \in [t_0, T + \theta].$$

Recall now that $(T, \psi(T)) \in \partial G_i$. This implies that we can find a pair $(T_1, x_1) \in {}^c G_i$ such that

$$|T - T_1| < \theta, \ |x_1 - \psi(T)| < r_1 \text{ and } |\psi(T) - \psi(T_1)| < r_1,$$

so that

$$|x_1 - \tilde{\varphi}(T_1)| < 4r_1.$$

Applying Theorem 1.5 we conclude that there exists a solution $\varphi(t)$ of (4.62) such that $\varphi(T_1) = x_1$ (hence $(T_1, \varphi(T_1)) \notin G_i$) and

$$|\varphi(t_0) - \tilde{\varphi}(t_0)| \leq 4r_1 e^{\int_{t_0}^{T_1} L(\tau)\, d\tau} \leq 4r_1 e^{\int_{t_0}^{T+1} L(\tau)\, d\tau} = \frac{r_0}{2}.$$

152   Differential inclusions

This yields $|\varphi(t_0) - x_0| \leq |\varphi(t_0) - \tilde{\varphi}(t_0)| + |\tilde{\varphi}(t_0) - x_0| < r_0$, which implies $(t_0, \varphi(t_0)) \in G_i$, a contradiction to the robust stability of (4.62). ∎

Propositions 4.7 and 4.8 will be used in next subsection.

4.2.2 Converse of first Liapunov theorem

Here, we assume that the origin is robustly stable and we wonder whether Theorem 4.6 may be inverted, i.e., if there exists a (continuous) weak Liapunov function $V(t, x)$ fulfilling (4.59) and (4.60). As it has been shown in Subsection 2.2.2, a smooth autonomous stable ODE (respectively, differential inclusion) may fail to possess a continuous (weak) Liapunov function of x alone. Moreover the existence of a continuous weak Liapunov function does not guarantee the existence of a smooth one. Therefore, in what follows we focus on the existence of a weak Liapunov function $V(t, x)$ (with the best regularity as possible) for (4.1). Proposition 4.3 tells us that some smoothness with respect to time is required for the existence of a weak Liapunov function which is *smooth*. The best result we get in general is as follows.

Theorem 4.7 *[16] Let $F : [0, +\infty) \times B_{h_0} \to \mathbb{R}^n$ be a set-valued map fulfilling ($\mathbf{H}_1$), ($\mathbf{H}_2$), ($\mathbf{H}_3$) and ($\mathbf{H}_4$). If the origin is robustly stable, then there exists a weak Liapunov function $V : [0, +\infty) \times B_h \to \mathbb{R}^+$ (h is some number $< h_0$) which is locally Lipschitz continuous.*

The method of proof is basically inspired by Kurzweil's ideas ([90], [91], [92]); however, a number of new nontrivial problems due to the multivalued nature of the right hand-side of the equation need to be dealt with.

Another converse of the first Liapunov theorem has been obtained for locally Lipschitz continuous set-valued maps, which assume compact (but not necessary convex) values.

Theorem 4.8 *[6] Let $F : [0, +\infty) \times B_{h_0} \to \mathbb{R}^n$ be a set-valued map which is compact valued, measurable with respect to t and locally Lipschitz continuous with respect to x. If the origin is uniformly stable for (4.1), then there exists a weak Liapunov function $V : [0, +\infty) \times B_h \to \mathbb{R}^+$ (h is some number $< h_0$) which is locally Lipschitz continuous.*

In contrast to Theorem 4.7, Theorem 4.8 may be proved by means of the following (explicit) Liapunov function candidate

$$V(t, x) := \inf_{\varphi} \inf_{s \leq t} \|\varphi(s)\|,$$

where φ ranges over the set of solutions of (4.1) such that $\varphi(t) = x$ and s belongs to the domain of φ. The details of the proof may be found in [6], or in the second step of the proof of Theorem 4.9. (See below Subsection 4.2.3.) Alternatively, Theorem 4.8 may be obtained as a direct consequence of Theorem 4.7 together with Propositions 4.7 and 4.8.

The last converse first Liapunov theorem presented in this section is concerned with the existence of a weak Liapunov function which is *smooth*.

Theorem 4.9 *[118]* (CONVERSE OF FIRST LIAPUNOV THEOREM) *Assume that (4.1) is robustly locally stable at the origin (respectively, robustly Lagrange stable) and that* (**H**′) *holds true. Then there exists a function V of class C^∞ on $\mathbb{R}^+ \times \mathbb{R}^n$ and two continuous functions $a, b \in \mathcal{K}_0^\infty$ such that, for some number $h > 0$,*

$$a(\|x\|) \leq V(t,x) \leq b(\|x\|), \quad \begin{array}{l} \forall t \geq 0, \ \|x\| < h \\ (\text{respectively}, \ \|x\| > h), \end{array} \quad (4.64)$$

$$\frac{\partial V}{\partial t}(t,x) + \langle \nabla_x V(t,x), v \rangle \leq 0, \quad \begin{array}{l} \forall t \in \mathbb{R}^+ \setminus N_0, \ \|x\| < h \\ (\text{respectively}, \ \|x\| > h) \\ \text{and } v \in F(t,x). \end{array} \quad (4.65)$$

Clearly, (4.65) implies that V is a weak Liapunov function for (4.1). The proof of Theorem 4.9 may be carried out by following the same pattern as for Theorem 4.5, the main change being in the definition of a locally Lipschitz continuous (weak) Liapunov function (second step). The proof (for the robust stability part) is reported in next section. It is shorter than the original one in [118].

4.2.3 Proof of the converse of first Liapunov theorem

The set E and the multivalued functions F_1 and F_2 are defined as in 4.1.4. We emphasize, however, that the set U and the functions $\delta(t,x)$ and $V(t,x)$ will be defined in another way.

FIRST STEP. (Regularization of F)

Once again, we extend F on $[-1, 0) \times \mathbb{R}^n$ by setting $F(t,x) := \{-x\}$, so that the origin is robustly locally stable for $t \geq -1$. Moreover, (**H**′) is fulfilled by F on $[-1, +\infty) \times \mathbb{R}^n$, provided that $0 \in N_0$. Let $(a_i)_{i \geq 0}$, $(b_i)_{i \geq 0}$ and $(G_i)_{i \geq 0}$ be as in Definition 3.12. (Here, we assume that $[-1, +\infty) \times \overline{B}_{a_i} \subset G_i \subset [-1, +\infty) \times B_{b_i}$.) The following lemma will be used in place of Lemma 4.3.

Lemma 4.10 *Let $i \geq 0$, $k \geq -1$ and let $K \subset ([k, k+1] \times \mathbb{R}^n) \cap G_i$ be any nonempty compact set. Then there exist two numbers $\delta > 0$, $\alpha > 0$ fulfilling the following property: for every $x(\cdot) : [t_1, t_2] \subset [k, k+1] \to \mathbb{R}^n$ such that*

- *$x(\cdot)$ is absolutely continuous,*
- *$(t_1, x(t_1)) \in K$,*
- *$\dot{x}(t) \in \overline{\text{co}}\{F(\overline{B}_\delta(t, x(t)) \cap E)\}$ for a.e. $t \in [t_1, t_2]$,*

we have
$$\min_{t \in [t_1, t_2]} \text{dist}\big((t, x(t)), {}^cG_i\big) \geq \alpha.$$

Proof. We argue by contradiction. If the statement is false, then there exist four sequences of numbers $(\delta^j)_{j \geq 0}$, $(\alpha^j)_{j \geq 0}$, $(t_1^j)_{j \geq 0}$ and $(t_2^j)_{j \geq 0}$ with $\delta^j \to 0$, $\alpha^j \to 0$, $k \leq t_1^j \leq t_2^j \leq k+1$ and a sequence $(x^j)_{j \geq 0}$ of absolutely continuous functions, $x^j(\cdot) : [t_1^j, t_2^j] \to \mathbb{R}^n$, such that for all $j \geq 0$, $(t_1^j, x^j(t_1^j)) \in K$, $\dot{x}^j(t) \in \overline{\text{co}}\{F(\overline{B}_{\delta^j}(t, x(t)) \cap E)\}$ for a.e. $t \in [t_1^j, t_2^j]$ and $\min_{t \in [t_1^j, t_2^j]} \text{dist}\big((t, x^j(t)), {}^cG_i\big) < \alpha^j$. Replacing t_2^j by some time $\tau^j \in [t_1^j, t_2^j]$ if needed, we may assume that the minimum of the distance of $(t, x^j(t))$ to ${}^cG_i$ is achieved for $t = t_2^j$. By Lemma 4.2, extracting subsequences, we may also assume that $(t_1^j, x^j(t_1^j)) \to (t_1, x(t_1))$, $(t_2^j, x^j(t_2^j)) \to (t_2, x(t_2))$ for some numbers $t_1, t_2 \in [k, k+1]$ and some solution $x(\cdot)$ of (4.1) on $[t_1, t_2]$. It follows that $(t_1, x(t_1)) \in K \subset G_i$ and that $\text{dist}\big((t_2, x(t_2)), {}^cG_i\big) = 0$, i.e., $(t_2, x(t_2)) \in {}^cG_i$, contradicting the positive invariance of G_i. ∎

We are in position to define the function $\delta(t, x)$. Let $i \geq 0$ be fixed. We define a sequence $\{K_i^k\}_{k \geq -1}$ of compact sets and two sequences $(\delta_i^k)_{k \geq -1}$, $(\alpha_i^k)_{k \geq -1}$ of numbers in an inductive way. For $k = -1$, we set $K_i^{-1} = [-1, 0] \times \overline{B}_{a_i}$, $\delta_i^{-1} = \delta$ and $\alpha_i^{-1} = \alpha$, where δ and α are given in Lemma 4.10, applied with $k = -1$, $K = K_i^{-1}$. Then, assume that K_i^l, δ_i^l and α_i^l have been defined for $-1 \leq l < k$. Set

$$K_i^k := [k, k+1] \times \overline{B}_{a_i} \cup \{(t, x): \ t = k \text{ and } \text{dist}\big((t, x), {}^cG_i\big) \geq \alpha_i^{k-1}\},$$

and set finally $\delta_i^k = \delta$, $\alpha_i^k = \alpha$, where δ and α are as given in Lemma 4.10 (with $K = K_i^k$).

The function $\delta : [-1, +\infty) \times \mathbb{R}^n \to \mathbb{R}^+$ is chosen in such a way that

$$\delta(t, x) = 0 \iff x = 0, \tag{4.66}$$

$$\delta(t, x) < \delta_i^k \text{ whenever } k \leq t < k+1 \text{ and } (t, x) \in G_i \setminus G_{i-1} \ (i \geq 0), \tag{4.67}$$

and δ is Lipschitz continuous with 1 as a Lipschitz constant. Once again, F_1 is defined as $F_1(t, x) := \overline{\text{co}}\{F(\overline{B}_{\delta(t, x)}(t, x) \cap E)\}$ for any $(t, x) \in [-1, +\infty) \times \mathbb{R}^n$.

To prove that the origin is still robustly stable for the system

$$\dot{x} \in F_1(t,x), \quad t \geq -1, \ x \in \mathbb{R}^n, \tag{4.68}$$

we need the following lemma (in place of Lemma 4.4).

Lemma 4.11 *Let $i \geq 0$, $(t_0, x_0) \in [-1, +\infty) \times \overline{B}_{a_i}$. Pick any solution $x(\cdot)$ of (4.68) such that $x(t_0) = x_0$. Then $(t, x(t)) \in G_i$ for all $t \geq t_0$.*

Proof. Otherwise, we may find a solution $x(\cdot)$ of (4.68) issuing from x_0 at $t = t_0$, and such that $(t_1, x(t_1)) \notin G_i$ for some $t_1 > t_0$. Hence, there exist two times t_2, t_3 with $t_0 \leq t_2 < t_3 \leq t_1$ and such that $\|x(t_2)\| = a_i$, $(t, x(t)) \in G_i \setminus [-1, +\infty) \times \overline{B}_{a_i}$ for $t_2 < t \leq t_3$ and $(t_3, x(t_3)) \notin G_i$. Let $k_0 := [t_2] \in \mathbb{N} \cup \{-1\}$. Clearly, $(t_2, x(t_2)) \in K_i^{k_0}$ and $(t, x(t)) \in G_i \setminus G_{i-1}$ for all $t \in [t_2, \min(t_3, k_0+1)]$. It follows from (4.68) that

$$\dot{x}(t) \in \overline{\text{co}}\{F(B_{\delta_i^k}(t, x(t)) \cap E)\} \quad \text{for a.e. } t \in [t_2, \min(t_3, k_0+1)],$$

hence, by virtue of the definition of $\delta_i^{k_0}$ and $\alpha_i^{k_0}$, $\text{dist}((t, x(t)), {}^c G_i) \geq \alpha_i^{k_0} > 0$ for $t_2 \leq t \leq \min(t_3, k_0 + 1)$, whereas $\text{dist}((t_3, x(t_3)), {}^c G_i) = 0$. We infer that $t_3 > k_0 + 1$. After a straightforward induction argument, we get $\min_{t \in [k, k+1]} \text{dist}((t, x(t)), {}^c G_i) \geq \alpha_i^k > 0$ for any $k < [t_3]$, and finally $\min_{t \in [[t_3], [t_3]+1]} \text{dist}((t, x(t)), {}^c G_i) \geq \alpha_i^{[t_3]} > 0$, contradicting $(t_3, x(t_3)) \in {}^c G_i$. ∎

Corollary 4.4 *The origin is locally robustly stable for (4.68).*

This result will be obtained as a by-product of next section, where a continuous weak Liapunov function $V(t, x)$ will be constructed.

We set

$$U := (-1, +\infty) \times (B_{b_0} \setminus \{0\}). \tag{4.69}$$

Let $W(t, x)$ be given by (4.22). The partition of unity $(\psi_i)_{i \geq 0}$ on U, and the multivalued map $F_2 : (-1, +\infty) \times B_{b_0} \to \mathbb{R}^n$ are defined as in the proof of the converse of second Liapunov theorem (Subsection 4.1.4).

SECOND STEP. (Construction of a locally Lipschitz continuous weak Liapunov function)

From now on, for any $(t_0, x_0) \in (-1, +\infty) \times B_{b_0}$ we denote by S_{t_0, x_0} the set of all the solutions $x(\cdot)$ of (4.20) such that $x(t_0) = x_0$, and by $\mathrm{dom}(x)$ the domain ($\subset (-1, +\infty)$) of $x(\cdot)$. We define the weak Liapunov function candidate as

$$V(t,x) := \inf_{\varphi \in \mathcal{S}_{t,x}} \inf_{s \in [-\frac{1}{2},t] \cap \mathrm{dom}(\varphi)} \|\varphi(s)\|, \quad (t,x) \in (-\frac{1}{2}, +\infty) \times B_{b_0}. \quad (4.70)$$

Clearly, $0 \leq V(t,x) \leq \|x\|$. The following result shows that V is positive definite.

Lemma 4.12 *Let $a(\cdot): [0, b_0] \to [0, a_1]$ be any continuous increasing function fulfilling $a(b_i) = a_{i+1}$ for all $i \geq 0$. Then $V(t,x) \geq a(\|x\|)$ for all $(t,x) \in (-\frac{1}{2}, +\infty) \times B_{b_0}$.*

Proof. Argue by contradiction, and assume that $V(t,x) < a(\|x\|)$ for some pair $(t,x) \in (-\frac{1}{2}, +\infty) \times B_{b_0}$. Then there exists $\varphi \in \mathcal{S}_{t,x}$ and $s \in [-\frac{1}{2}, t] \cap \mathrm{dom}(\varphi)$ such that $V(t,x) \leq \|\varphi(s)\| < a(\|x\|)$. Let $i \in \mathbb{N}$ be such that $b_{i+1} < \|x\| \leq b_i$. It follows that $\|\varphi(s)\| < a(\|x\|) \leq a(b_i) = a_{i+1}$. Applying Lemma 4.11, we infer that $(t,x) = (t, \varphi(t)) \in G_{i+1}$, hence $\|x\| < b_{i+1}$, a contradiction. ∎

We now check that V is nonincreasing along the trajectories of (4.20).

Lemma 4.13 *Let $(t_0, x_0) \in (-\frac{1}{2}, +\infty) \times B_{b_0}$ and $\varphi \in \mathcal{S}_{t_0, x_0}$. Then $V(t_1, \varphi(t_1)) \leq V(t_0, x_0)$ for all $t_1 \geq t_0$.*

Proof. Pick any $t_1 \geq t_0$ and any $\psi \in \mathcal{S}_{t_0, x_0}$. Then the function

$$\tilde{\varphi}(t) := \begin{cases} \psi(t) & \text{if } t \in \mathrm{dom}(\psi) \cap (-\infty, t_0], \\ \varphi(t) & \text{if } t \in \mathrm{dom}(\varphi) \cap [t_0, +\infty), \end{cases}$$

belongs to $\mathcal{S}_{t_1, \varphi(t_1)}$, hence $\|\psi(s)\| = \|\tilde{\varphi}(s)\| \geq V(t_1, \varphi(t_1))$ for each $s \in [-\frac{1}{2}, t_0] \cap \mathrm{dom}(\psi)$. Taking the definition of $V(t_0, x_0)$ into account, we get $V(t_0, x_0) \geq V(t_1, \varphi(t_1))$, as wanted. ∎

It remains to prove that V is locally Lipschitz continuous on $(-\frac{1}{2}, +\infty) \times (B_{a_1} \setminus \{0\})$.

Proposition 4.9 *Let $\bar{t} > -\frac{1}{2}$ and $\bar{x} \in B_{a_1} \setminus \{0\}$. Then there exist three positive numbers δ, η, L such that for all pairs (t_1, x_1), (t_2, x_2) in $(\bar{t} - \eta, \bar{t} + \eta) \times B_\delta(\bar{x})$, we have*

$$|V(t_1, x_1) - V(t_2, x_2)| \leq L(|t_1 - t_2| + \|x_1 - x_2\|). \quad (4.71)$$

Proof. Fix $\bar{t}$, $\bar{x}$ as in the statement and let $i \geq 1$ be such that

$$(b_{i+2} <) a_{i+1} \leq \|\bar{x}\| < a_i. \quad (4.72)$$

F_2 being locally Lipschitz continuous on U, there exists a constant $K > 0$ such that for any pairs (t_1, x_1), (t_2, x_2) with $t_1, t_2 \in [-\frac{1}{2}, \bar{t}+1]$, $\|x_1\|, \|x_2\| \in [\frac{a_{i+3}}{2}, a_{i-1}]$,

$$h\big(F_2(t_1,x_1), F_2(t_2,x_2)\big) \leq K\big(|t_1-t_2| + \|x_1-x_2\|\big). \qquad (4.73)$$

Let $M > 0$ be such that for each pair $(t,x) \in [-\frac{1}{2}, \bar{t}+1] \times \overline{B}_{b_1}$,

$$F_2(t,x) \subset \overline{B}_M. \qquad (4.74)$$

Pick finally $\delta > 0$ such that

$$2\delta < \min\left(a_i - \|\bar{x}\|, \|\bar{x}\| - b_{i+2}, \frac{e^{-K(\bar{t}+\frac{3}{2})}}{2} \min(\frac{a_{i+3}}{2}, a_{i-1} - b_i)\right). \qquad (4.75)$$

and $\eta \in \big(0, \min(\bar{t}+\frac{1}{2}, \frac{\delta}{2M}, 1)\big)$.

We shall need the following result:

Lemma 4.14 *Let $t_1 \in [\bar{t}-\eta, \bar{t}+\eta]$ and $x_1 \in B_{2\delta}(\bar{x})$ be given. Let $\varphi_1 \in \mathcal{S}_{t_1,x_1}$ be defined on some interval $[\tau, t_1] \subset [-\frac{1}{2}, \bar{t}+\eta]$ and assume that $\|\varphi_1(\tau)\| < a_i$. Let $x_2 \in B_{2\delta}(\bar{x})$. Then there exists a trajectory $\varphi_2 \in \mathcal{S}_{t_1,x_2}$ defined on $[\tau, t_1]$, such that*

$$\forall t \in [\tau, t_1], \quad \|\varphi_1(t) - \varphi_2(t)\| \leq \|x_1 - x_2\| \, e^{K|t-t_1|}. \qquad (4.76)$$

Proof. Since $\|\varphi_1(\tau)\| < a_i$, we have $\|\varphi_1(t)\| < b_i$ for any $t \in [\tau, t_1]$. On the other hand, for any $t \in [\tau, t_1]$,

$$\|\varphi_1(t)\| \geq V(t_1, x_1) \geq a(\|x_1\|) > a(\|\bar{x}\| - 2\delta) > a(b_{i+2}) = a_{i+3}, \qquad (4.77)$$

since, by (4.75), $b_{i+2} < \|\bar{x}\| - 2\delta$. Let $b := 4\delta \, e^{K(\bar{t}+\frac{3}{2})} < \min(\frac{a_{i+3}}{2}, a_{i-1}-b_i)$. Since the domain $t \in [\tau, t_1]$, $\|x - \varphi_1(t)\| < b$ is contained in the domain $t \in [-\frac{1}{2}, \bar{t}+1]$, $\|x\| \in [\frac{a_{i+3}}{2}, a_{i-1}]$ (where F_2 is Lipschitz continuous with Lipschitz constant K), it follows from Lemma 4.1 that there exists a solution $\varphi_2 \in \mathcal{S}_{t_1,x_2}$ (defined on the left of t_1), such that $\|\varphi_1(t) - \varphi_2(t)\| \leq \|x_1 - x_2\| e^{K|t-t_1|}$ as long as $\|x_1 - x_2\| e^{K|t-t_1|} \leq b$. Since $\|x_1 - x_2\| < 4\delta$ and $|\tau - t_1| \leq \bar{t}+\frac{3}{2}$, we conclude that (4.76) holds true (at least) on $[\tau, t_1]$, as desired. ∎

We go back to the proof of Proposition 4.9. Let $(t_1, x_1), (t_2, x_2) \in (\bar{t}-\eta, \bar{t}+\eta) \times B_\delta(\bar{x})$. Without loss of generality, we may assume that $t_1 \leq t_2$. We need some preliminary steps.

Step 1. $|V(t_2, x_1) - V(t_2, x_2)| \leq \|x_1 - x_2\| e^{K(\bar{t}+\frac{3}{2})}$.

Proof. By definition of $V(t_2, x_1)$, for each (small) $\sigma > 0$ there exists a solution $\varphi_1 \in \mathcal{S}_{t_2, x_1}$ and $\tau \in \mathrm{dom}(\varphi_1) \cap [-\frac{1}{2}, t_2]$ such that

$$V(t_2, x_1) \leq \|\varphi_1(\tau)\| < V(t_2, x_1) + \sigma \leq \|x_1\| + \sigma < a_i, \tag{4.78}$$

hence

$$V(t_2, x_2) - V(t_2, x_1) < V(t_2, x_2) - \|\varphi_1(\tau)\| + \sigma. \tag{4.79}$$

By Lemma 4.14, there exists a solution $\varphi_2 \in \mathcal{S}_{t_2, x_2}$ such that, for all $t \in [\tau, t_2]$,

$$\|\varphi_1(t) - \varphi_2(t)\| \leq \|x_1 - x_2\| e^{K|t-t_2|} \leq \|x_1 - x_2\| e^{K(\bar{t}+\frac{3}{2})}. \tag{4.80}$$

Using the definition of $V(t_2, x_2)$, we get $V(t_2, x_2) \leq \|\varphi_2(\tau)\|$, hence, using (4.79) and (4.80),

$$\begin{aligned} V(t_2, x_2) - V(t_2, x_1) &< \|\varphi_2(\tau)\| - \|\varphi_1(\tau)\| + \sigma \\ &\leq \|\varphi_2(\tau) - \varphi_1(\tau)\| + \sigma \\ &\leq \|x_1 - x_2\| e^{K(\bar{t}+\frac{3}{2})} + \sigma. \end{aligned}$$

The reasoning can be repeated by exchanging the roles of x_1 and x_2. The proof of Step 1 is completed by observing that σ is arbitrary.

<u>Step 2.</u> $V(t_2, x_1) - V(t_1, x_1) \leq M e^{K(\bar{t}+\frac{3}{2})} |t_1 - t_2|$.
Proof. Pick any $\varphi_1 \in \mathcal{S}_{t_1, x_1}$ and set $\tilde{x}_2 = \varphi_1(t_2)$. Since $V(t, \varphi_1(t))$ is nonincreasing,

$$V(t_2, x_1) - V(t_1, x_1) \leq V(t_2, x_1) - V(t_2, \tilde{x}_2).$$

On the other hand

$$\|x_1 - \tilde{x}_2\| \leq \int_{t_1}^{t_2} \|\dot{\varphi}_1(\tau)\| d\tau \leq M |t_2 - t_1| \leq 2M\eta < \delta,$$

hence $\tilde{x}_2 \in B_{2\delta}(\bar{x})$. By repeating the same argument as in Step 1, we get

$$|V(t_2, x_1) - V(t_2, \tilde{x}_2)| \leq \|x_1 - \tilde{x}_2\| e^{K(\bar{t}+\frac{3}{2})}$$

hence

$$V(t_2, x_1) - V(t_1, x_1) \leq M e^{K(\bar{t}+\frac{3}{2})} |t_1 - t_2|.$$

<u>Step 3.</u> $V(t_1, x_1) - V(t_2, x_1) \leq M e^{K(\bar{t}+\frac{3}{2})} |t_1 - t_2|$.
By the definition of $V(t_2, x_1)$, for each $\sigma > 0$ there exists a solution $\psi \in \mathcal{S}_{t_2, x_1}$ and a time $\tau \in \mathrm{dom}(\psi) \cap [-\frac{1}{2}, t_2]$ such that

$$V(t_2, x_1) \leq \|\psi(\tau)\| < V(t_2, x_1) + \sigma < a_i,$$

hence
$$V(t_1, x_1) - V(t_2, x_1) \le V(t_1, x_1) - \|\psi(\tau)\| + \sigma.$$

Let us distinguish two cases.

First case: $\tau \le t_1$.
Let $x_3 = \psi(t_1)$. Then $\|x_3 - x_1\| \le \int_{t_1}^{t_2} \|\dot\psi(\tau)\| d\tau \le |t_2 - t_1| M < \delta$, hence $x_3 \in B_{2\delta}(\bar x)$. It follows from Lemma 4.14 that there exists a solution $\varphi \in \mathcal{S}_{t_1,x_1}$ defined on $[\tau, t_1]$, and such that for all $t \in [\tau, t_1]$
$$\|\varphi(t) - \psi(t)\| \le \|x_1 - x_3\| e^{K|t-t_1|}.$$

Since $V(t_1, x_1) \le \|\varphi(\tau)\|$, we have
$$\begin{aligned}V(t_1, x_1) - V(t_2, x_1) &\le \|\varphi(\tau)\| - \|\psi(\tau)\| + \sigma \\ &\le \|\varphi(\tau) - \psi(\tau)\| + \sigma \\ &\le \|x_1 - x_3\| e^{K(\bar t + \frac{3}{2})} + \sigma \\ &\le M e^{K(\bar t + \frac{3}{2})} |t_2 - t_1| + \sigma.\end{aligned}$$

Second case: $t_1 < \tau \le t_2$.
Since $V(t_1, x_1) \le \|x_1\|$, we have
$$\begin{aligned}V(t_1, x_1) - V(t_2, x_1) &\le \|x_1\| - \|\psi(\tau)\| + \sigma \\ &\le \int_\tau^{t_2} \|\dot\psi(\tau)\| d\tau + \sigma \\ &\le M |t_2 - t_1| + \sigma \\ &\le M e^{K(\bar t + \frac{3}{2})} |t_2 - t_1| + \sigma.\end{aligned}$$

Since the choice of σ was arbitrary we conclude that
$$V(t_1, x_1) - V(t_2, x_1) \le M e^{K(\bar t + \frac{3}{2})} |t_2 - t_1|,$$

as required.

CONCLUSION. Using Steps 1, 2 and 3, we get
$$\begin{aligned}|V(t_1, x_1) - V(t_2, x_2)| &\le |V(t_1, x_1) - V(t_2, x_1)| + |V(t_2, x_1) - V(t_2, x_2)| \\ &\le M e^{K(\bar t + \frac{3}{2})} |t_2 - t_1| + e^{K(\bar t + \frac{3}{2})} \|x_2 - x_1\| \\ &\le L(\|x_1 - x_2\| + |t_1 - t_2|)\end{aligned}$$

where $L := \max(1, M) e^{K(\bar t + \frac{3}{2})}$. The proof of Proposition 4.9 is completed. ∎

Corollary 4.5 *For a.e. $(t_0, x_0) \in (-\frac{1}{2}, +\infty) \times B_{a_1}$,*
$$\forall v \in F_2(t_0, x_0), \quad \frac{\partial V}{\partial t}(t_0, x_0) + \langle \nabla_x V(t_0, x_0), v \rangle \le 0. \tag{4.81}$$

The proof is the same as for Corollary 4.3.

THIRD STEP. (Regularization of V)

To smooth V, we need to strengthen slightly the inequality in (4.81). This can be done by using (as in [118]) some trick due to Yoshizawa [159]. Set, for $(t, x) \in (-\frac{1}{2}, +\infty) \times B_{a_1}$

$$V_L(t, x) := \frac{1+t}{1+2t} V(t, x) \quad \text{and} \quad W_L(t, x) := \frac{1}{(1+2t)^2} a(\|x\|).$$

Then $\frac{1}{2} a(\|x\|) \leq V_L(t, x) \leq \|x\|$ for each (t, x) and V_L is locally Lipschitz continuous on $(-\frac{1}{2}, +\infty) \times (B_{a_1} \setminus \{0\})$. On the other hand, it follows from Corollary 4.5 that for a.e. $(t_0, x_0) \in (-\frac{1}{2}, +\infty) \times (B_{a_1} \setminus \{0\})$,

$$\forall v \in F_2(t_0, x_0), \quad \frac{\partial V_L}{\partial t}(t_0, x_0) + \langle \nabla_x V_L(t_0, x_0), v \rangle \leq -W_L(t_0, x_0). \quad (4.82)$$

Next lemma is proved along the same lines as for Lemma 4.9.

Lemma 4.15 *Let S be any compact set in $\tilde{U} := (-\frac{1}{2}, +\infty) \times (B_{a_1} \setminus \{0\})$ and let $\varepsilon > 0$. Then there exists a function $\overline{V}$ of class C^∞ and with compact support in $(-\frac{1}{2}, +\infty) \times B_{a_1}$, such that*

$$\|\overline{V} - V\|_{L^\infty(S)} < \varepsilon$$

and

$$\forall (t_0, x_0) \in S, \ \forall v \in F_2(t_0, x_0), \quad \frac{\partial \overline{V}}{\partial t}(t_0, x_0) + \langle \nabla_x \overline{V}(t_0, x_0), v \rangle \leq -\frac{1}{2} W_L(t_0, x_0).$$

With this lemma at hand we may smooth V_L in exactly the same way as in the proof of Theorem 4.5. This completes the proof of Theorem 4.9. ∎

4.2.4 Application to external stabilization

Let us now consider the following control system:

$$\dot{x} = f(t, x, u), \quad (4.83)$$

where $t \geq 0$ is time, $x \in \mathbb{R}^n$ is the state variable and $u \in \mathbb{R}^m$ is the control variable. We aim to investigate the links between the (robust) Lagrange stability of the unforced system

$$\dot{x} = f(t, x, 0) \quad (4.84)$$

and the UBIBS stability of (4.83). Assume that f is affine with respect to u. Then a result in the same vein as Theorem 2.25 may be proved thanks to Theorem 4.9.

Theorem 4.10 *[118] Consider a system*

$$\dot{x} = f_0(t,x) + \sum_{i=1}^m u_i\, f_i(t,x) \qquad (4.85)$$

where for each $i \in \{0,\ldots,m\}$ the field $f_i \in \mathcal{L}^\infty_{loc}(\mathbb{R}^+ \times \mathbb{R}^n, \mathbb{R}^n)$ and it is ECT. Assume that the unforced system $\dot{x} = f_0(t,x)$ is (robustly) Lagrange stable. Then the system (4.85) is UBIBS stabilizable (in Filippov's sense) by means of a feedback law $u = k(t,x) + v$ such that $k \in \mathcal{L}^\infty_{loc}(\mathbb{R}^+ \times \mathbb{R}^n, \mathbb{R}^m)$ and k is ECT, as well. Moreover, if for each $i \geq 0$ f_i is of class C^r for some $r \in \mathbb{N} \cup \{\infty\}$, then k is also of class C^r.

Notice that here we only assume the (robust) Lagrange stability of the unforced system.

4.3 Nonsmooth Liapunov functions

In Subsection 2.3.1 we saw that, because of certain topological obstructions, the problem of the asymptotic stabilization of nonlinear systems leads in a natural way to deal with discontinuous feedback laws and hence, to systems of ordinary differential equations with a discontinuous right hand side. As remarked in Chapter 1, ordinary differential equations with discontinuous right hand side require a generalization of the usual definition of solution. The notion of Filippov solution is the most popular approach to this problem. As already recalled in Subsection 1.2.1, the first step of Filippov's idea consists in replacing the discontinuous differential system by a differential inclusion, whose right hand side can be proven to be convex and compact valued, and upper semi-continuous. By virtue of Theorem 4.3, it is therefore clear that in principle, for analyzing the stability properties of a discontinuous differential equation with respect to Filippov solutions, it is sufficient to look for candidate Liapunov functions of class C^∞. However, for concrete problems the explicit construction of a Liapunov function with such a high degree of regularity may be a very hard or even impossible task: sometimes, one can take much more advantage of the use of a much less regular Liapunov function. In order to illustrate this statement, we discuss an example, which appears in [60] (with time reversed) and also in [152], [105].

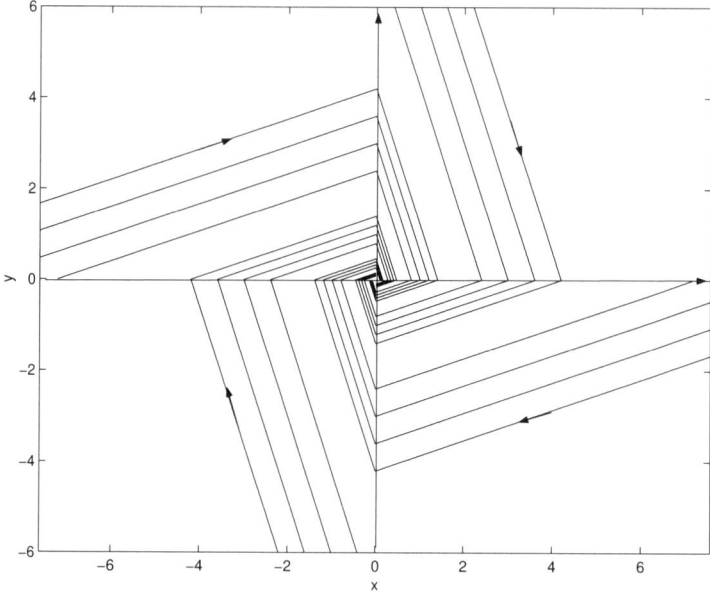

Figure 4.1: A discontinuous asymptotically stable system

Consider the two-dimensional, discontinuous differential system

$$\begin{cases} \dot{x} = -\mathrm{sgn}\, x + 2\mathrm{sgn}\, y \\ \dot{y} = -2\mathrm{sgn}\, x - \mathrm{sgn}\, y \ . \end{cases} \quad (4.86)$$

The discontinuities are situated at points where either $x = 0$ or $y = 0$, and are of a very simple type. In particular, we see that for each nonzero initial state there exists one and only one Filippov solution, and that the origin is globally asymptotically stable (see Figure 4.1; actually, the origin is reached in finite time from each initial state). We can imagine a candidate C^∞ Liapunov function for this system, as a function $V(x, y)$ whose level curves are drawn in Figure 4.2: we note in particular that the monotonicity condition is fulfilled since the trajectories of the system cross the level curves of $V(x, y)$ in the right way at every point (see Figure 4.3).

Even if we have now conceived a more or less clear idea of the geometric shape of the graph of $V(x, y)$, it is clear that we have a very little hope to obtain a useful analytic expression of it. On the contrary, the problem can be easily handled by resorting to the Lipschitz continuous, but not everywhere differentiable, candidate Liapunov function $V(x, y) = |x| + |y|$. The level sets of this function are plotted in Figure 4.4, and Figure 4.5 shows that even in

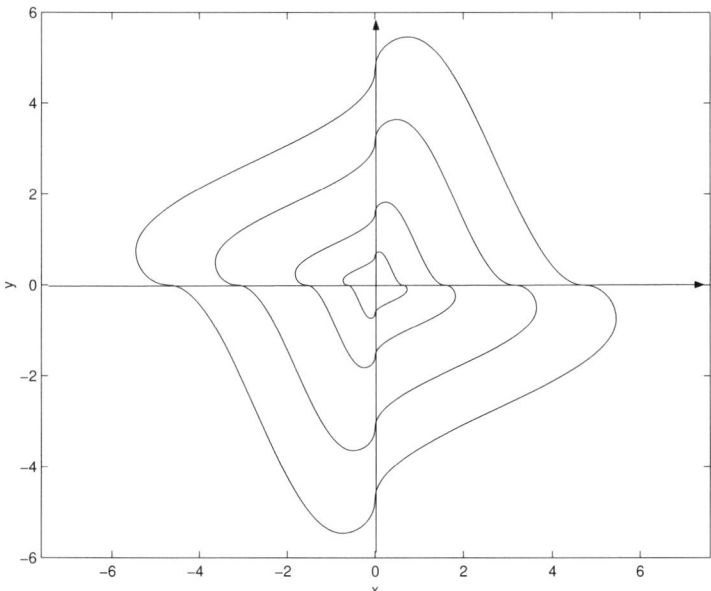

Figure 4.2: Level sets of a smooth candidate Liapunov function

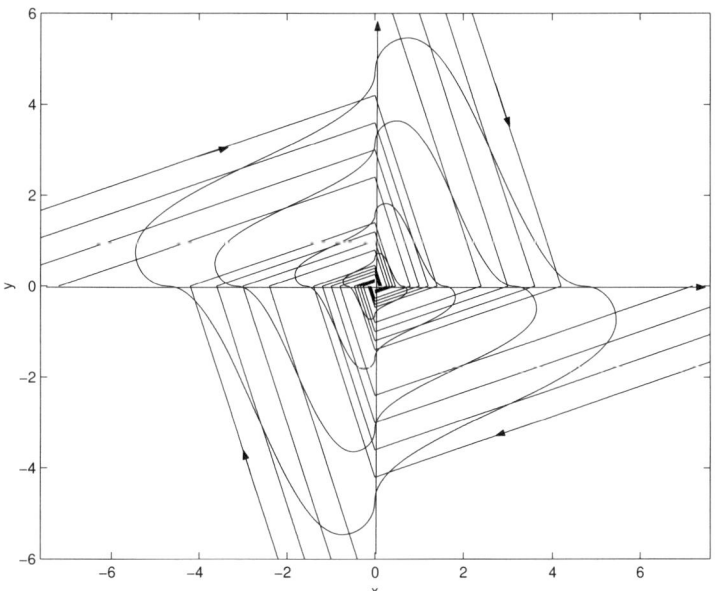

Figure 4.3: Level curves crossing

164   Differential inclusions

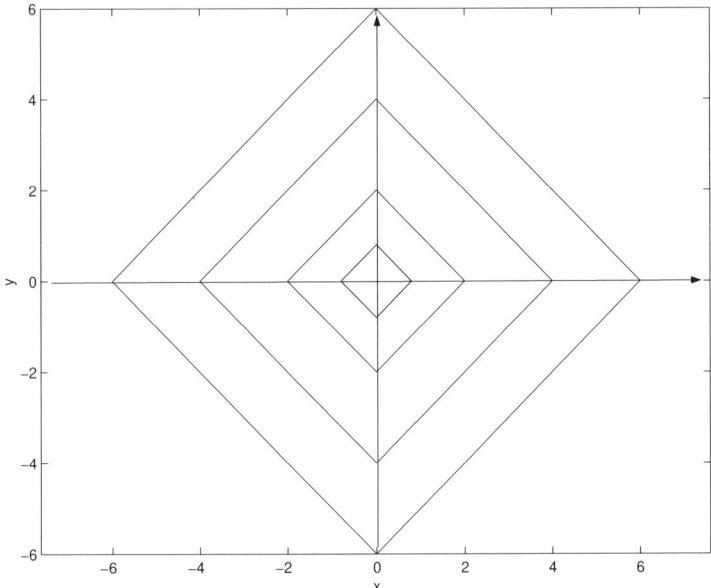

Figure 4.4: Level sets of a nonsmooth candidate Liapunov function

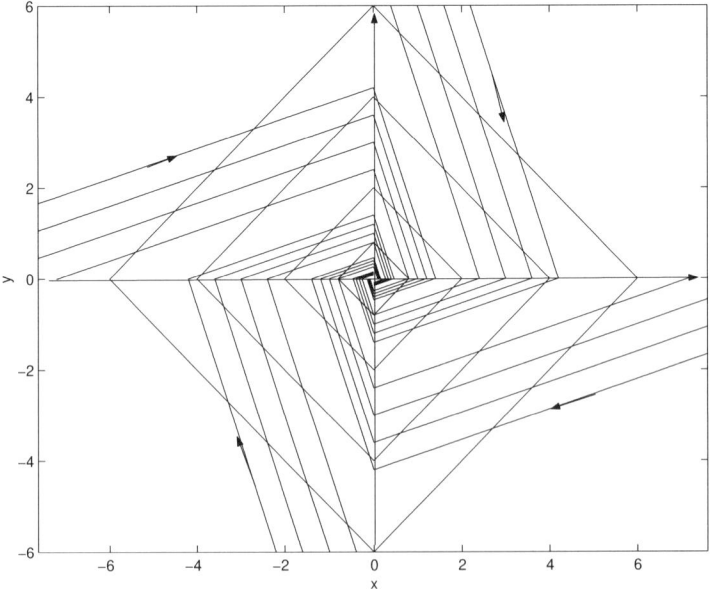

Figure 4.5: Level curves crossing

this case the trajectories of the system cross the level curves in such a way that the monotonicity condition is fulfilled. The natural objection is, of course, that at the points where the gradient of $V(x,y)$ does not exist, the monotonicity condition cannot be reduced to a differential inequality. As a matter of fact, the difficulty can be overcame by invoking some tools from nonsmooth analysis. This is actually the subject treated in Chapter 6.

Chapter 5

Additional properties of strict Liapunov functions

In this chapter, we go back to an ODE of the form

$$\dot{x} = f(x), \tag{5.1}$$

where f is a smooth vector field which is defined in a neighborhood of 0 and fulfills $f(0) = 0$. We shall assume throughout that the origin is locally asymptotically stable (AS, in short) for (5.1). In this chapter, $\|\cdot\|$ will denote the usual Euclidean norm, instead of the sup-norm.

To introduce the issues discussed here, let us consider the (trivial) situation where f is an analytic vector field in $\mathbb{R}$, i.e., $f(x) = a_0 + a_1 x + a_2 x^2 + \cdots$, for (say) $x \in (-r, r)$. Let $a_k x^k$ be the first not vanishing term in the series. Clearly, 0 is an AS equilibrium position for (5.1) if and only if

$$k \text{ is odd, and } a_k < 0. \tag{5.2}$$

(Therefore the asymptotic stability problem is algebraically solvable in the one-dimensional case.) On the other hand, if 0 is AS for (5.1), then

(i) $V(x) := x^2$ is a convenient strict Liapunov function for (5.1);

(ii) For any (smooth) vector field g such that $|g(x)| = O(|x|^{k+1})$ as $x \to 0$, the origin is still AS for the perturbed differential equation

$$\dot{x} = f(x) + g(x); \tag{5.3}$$

(iii) There exist positive constants δ, M_1, M_2 such that for any solution $x(\cdot)$ of (5.1) fulfilling $|x(0)| < \delta$,
(if $k = 1$)
$$M_1 e^{-|a_1|t}|x(0)| \leq |x(t)| \leq M_2 e^{-|a_1|t}|x(0)|, \tag{5.4}$$
(if $k > 1$)
$$M_1\bigl(1 + |x(0)|^{k-1}t\bigr)^{-\frac{1}{k-1}}|x(0)| \leq |x(t)| \leq M_2\bigl(1 + |x(0)|^{k-1}t\bigr)^{-\frac{1}{k-1}}|x(0)|. \tag{5.5}$$

The situation is not so trivial in higher dimension. First of all, for the system (5.1) to be AS at the origin it is *sufficient*, but not *necessary*, that this property already holds true in the first homogeneous approximation. A trivial example may be constructed in perturbing the harmonic oscillator.

Example 5.1 Consider the following perturbed harmonic oscillator.
$$\begin{cases} \dot{x}_1 = x_2 - x_1^3, \\ \dot{x}_2 = -x_1. \end{cases} \tag{5.6}$$

The first not vanishing homogeneous vector field in the Taylor expansion (namely, the linear part $f_1(x_1, x_2) = x_2 \frac{\partial}{\partial x_1} - x_1 \frac{\partial}{\partial x_2}$), gives rise to a system which is only *stable*, whereas the whole system is AS. This example shows that a whole segment of the Taylor series must be considered, in general, to know whether the asymptotic stability holds true for the whole system. ∎

On the other hand, if f is a polynomial vector field of degree k such that the origin is AS for (5.1), then the robustness property (ii) may fail to be true. This phenomenon occurs in the following example, discovered by the first author in 1985.

Example 5.2 Let $m \in \mathbb{R}$ be given, and consider the system
$$\begin{cases} \dot{x}_1 = -x_1^3 + x_2^3, \\ \dot{x}_2 = -x_1 + mx_2^5. \end{cases} \tag{5.7}$$

Then it is proved in [64] that the origin is a locally (respectively, globally) AS equilibrium position if, and only if, $m < 3/5$ (respectively, $m \leq 0$). Therefore, property (ii) is false for $k = 3$, $f := (-x_1^3 + x_2^3)\frac{\partial}{\partial x_1} - x_1 \frac{\partial}{\partial x_2}$, and $g := x_2^5 \frac{\partial}{\partial x_2}$. ∎

We now turn to the type of decay for the trajectories of (5.1), when f is a polynomial vector field of degree at most k, with $k > 1$. It should be tempting to conjecture that the solutions $x(\cdot)$ decrease at worse as in (iii), i.e.,

$$\|x(t)\| \leq Const.\bigl(1 + \|x(0)\|^{k-1}t\bigr)^{-\frac{1}{k-1}} \|x(0)\|. \tag{5.8}$$

Actually, solutions may decrease more slowly.

Example 5.2 (Continued) Consider system (5.7) with $m = 0$, i.e.,

$$\begin{cases} \dot{x}_1 &= -x_1^3 + x_2^3, \\ \dot{x}_2 &= -x_1. \end{cases} \tag{5.9}$$

(5.9) is an odd polynomial system. Hence, if $x(t)$ is the solution issuing from some x_0 at $t = 0$, then $-x(t)$ is the solution issuing from $-x_0$ at $t = 0$. Therefore, the analysis may be restricted to the half-plane $x_2 \geq 0$. Following [64], we define "polar" coordinates (r, θ) (with $r \geq 0$, $\theta \in [0, \pi]$) in the half-plane $x_2 \geq 0$ by

$$\begin{cases} x_1 &= \sqrt{2}\, r \cos \theta, \\ x_2^2 &= 2\, r \sin \theta. \end{cases}$$

Then $r^2 = x_1^2/2 + x_2^4/4$, and in the (r, θ) coordinates (5.9) becomes

$$\begin{cases} \dot{r} &= -2r^3 \cos^4 \theta, \\ \dot{\theta} &= -2\sqrt{r} \sin \theta \bigl(1 - r^{\frac{3}{2}} \cos^3 \theta \sin^{\frac{1}{2}} \theta\bigr). \end{cases} \tag{5.10}$$

Since $\dot{r} \geq -2r^3$, we get $r(t) \geq \bigl(r(0)^{-2} + 4t\bigr)^{-\frac{1}{2}}$ (even if $x_2(t) < 0$); hence, if $(t_n)_{n \geq 1}$ denotes the increasing sequence of times at which the coordinate x_1 vanishes, we obtain

$$|x_2(t_n)| = \sqrt{2r(t_n)} \geq \sqrt{2}\bigl(r(0)^{-2} + 4t_n\bigr)^{-\frac{1}{4}}.$$

Since (t_n) is unbounded, (5.8) cannot hold with $k - 3$. ■

We finish this introduction by pointing out that there exist algebraic AS systems without any *analytic* Liapunov function. (See below Subsection 5.2.2.) Sufficient conditions for the existence of a quadratic Liapunov function will be given in Subsection 5.1.1.

The chapter is outlined as follows. In the first section, we express the type of decay of the trajectories by means of some suitable Liapunov function. In the following section, we restrict ourselves to analytic systems and we investigate a few natural questions, e.g., the issue whether the asymptotic stability problem is (or is not) analytically solvable, the issue of the existence of some analytic Liapunov function, etc. In the third section we introduce the concept of weighted homogeneity. The key result is that a homogeneous AS system possesses a homogeneous Liapunov function. Four applications of this result to the stabilization problem are given. In the last section we extend the above development to more general symmetries.

5.1 Estimates for the convergence of trajectories

5.1.1 Exponential stability

Let $f : \Omega \subset \mathbb{R}^n \to \mathbb{R}^n$ be a vector field of class C^1, and let $A := \left(\frac{\partial f}{\partial x}\right)_{|x=0}$ be the Jacobian matrix of f at the origin. Recall that if A is Hurwitz (i.e., any eigenvalue of A has a negative real part), then the origin is locally AS for the linear system $\dot{x} = Ax$, and also for the nonlinear system (5.1). Moreover, the solutions decrease exponentially for both systems. Let us introduce some definition.

Definition 5.1 *The origin is said to be* exponentially stable *for (5.1) if there exist three numbers $\omega < 0$, $M > 0$ and $\delta > 0$ such that for any $x_0 \in B_\delta$, the solution $x(\cdot)$ of (5.1) issuing from x_0 at $t = 0$ is defined on $[0, +\infty)$ and it fulfills*

$$\forall t \geq 0, \quad \|x(t)\| \leq M e^{\omega t} \|x_0\|. \tag{5.11}$$

The infimum of the numbers $\omega < 0$ for which (5.11) is satisfied (for some constants $M, \delta > 0$) is called the exponent *of 0.*

Clearly, if A is Hurwitz, then 0 is exponentially stable for (5.1). It turns out that the converse is true, as well.

Theorem 5.1 *Let f be a vector field of class C^1 on a neighborhood Ω of 0 in $\mathbb{R}^n$, and assume that $f(0) = 0$. Then (5.1) is exponentially stable at 0 if and only if the Jacobian matrix $A = \left(\frac{\partial f}{\partial x}\right)_{|x=0}$ is Hurwitz. Moreover the exponent of 0 is $\sup\{\mathcal{R}e\,\lambda,\ \lambda \in \sigma(A)\}$.*

The reader is referred to [162] for a proof of this classical result. As a byproduct we obtain a useful characterization of exponential stability in terms of Liapunov functions.

Theorem 5.2 *Let f be a vector field of class C^1 near 0 and such that $f(0) = 0$. Then the following statements are equivalent.*
(i) *0 is exponentially stable for (5.1);*
(ii) *There exists a function V of class C^1 in a neighborhood of 0 such that, for some positive constants C_1, C_2, C_3, r and δ,*

$$\|x\| < \delta \quad \Rightarrow \quad C_1 \|x\|^r \leq V(x) \leq C_2 \|x\|^r, \tag{5.12}$$

$$\|x\| < \delta \quad \Rightarrow \quad \langle \nabla V(x), f(x) \rangle \leq -C_3 \|x\|^r; \tag{5.13}$$

(iii) *There exists a symmetric positive definite matrix $S \in \mathbb{R}^{n \times n}$ such that, for some positive constants C, δ,*

$$\|x\| < \delta \quad \Rightarrow \quad \langle Sx, f(x) \rangle \leq -C\|x\|^2. \tag{5.14}$$

Proof. (i)$\Rightarrow$(iii): According to Theorem 5.1, the matrix $A = \left(\frac{\partial f}{\partial x}\right)_{|x=0}$ is Hurwitz, hence there exists a symmetric, positive definite matrix $S \in \mathbb{R}^{n \times n}$ such that, for some constant $C' > 0$, $\langle Sx, Ax \rangle \leq -C'\|x\|^2$ for all $x \in \mathbb{R}^n$. Since $f(x) = Ax + o(\|x\|)$, (5.14) follows at once.
(iii)$\Rightarrow$(ii): It is sufficient to set $V(x) := \langle Sx, x \rangle$ and to take $r = 2$.
(ii)$\Rightarrow$(i): Pick any trajectory $x(\cdot)$ of (5.1) and set $v(t) := V(x(t))$. By (5.12) and (5.13)

$$\dot{v}(t) \leq -\frac{C_3}{C_2} v(t),$$

hence (using again (5.12))

$$\|x(t)\| \leq \left(\frac{v(0) e^{-\frac{C_3}{C_2} t}}{C_1} \right)^{\frac{1}{r}} \leq \left(\frac{C_2}{C_1} \right)^{\frac{1}{r}} e^{-\frac{C_3}{C_2 r} t} \|x(0)\|.$$

■

Remark 5.1 It follows from Theorem 5.2 that if the equilibrium position 0 is exponentially stable, then there exists a strict Liapunov function which is a *quadratic form*: $V(x) = \langle Sx, x \rangle$, $S > 0$. For AS systems, however, the corresponding strict Liapunov functions are in general not so simple (sometimes not even analytic: see below Subsection 5.2.2.).

5.1.2 Rational stability

When the Jacobian matrix A is no longer Hurwitz, then the solutions do not decrease exponentially. Instead, it may sometimes be proved that the solutions decrease like t^{-r}, $r > 0$.

Definition 5.2 *The origin is said to be* **rationally stable** *if there exist positive numbers M, k, η and δ (with $\eta \leq 1$) such that for any $x_0 \in B_\delta$, the solution $x(\cdot)$ of (5.1) issuing from x_0 at $t = 0$ is defined on $[0, +\infty)$ and it fulfills*

$$\forall t \geq 0, \quad \|x(t)\| \leq M\big(1 + \|x_0\|^k t\big)^{-\frac{1}{k}} \|x_0\|^\eta. \tag{5.15}$$

If V is any function defined in a neighborhood of 0, we set for all x_0

$$\dot{V}(x_0) := \overline{D^+ V(x(t))}_{|t=0} = \limsup_{t \to 0^+} \frac{V(x(t)) - V(x_0)}{t}.$$

The rational stability may be characterized by means of certain Liapunov functions [66, Theorem 23.1].

Theorem 5.3 *Let f be a vector field of class C^1 near 0 and such that $f(0) = 0$. Then the origin is rationally stable if and only if there exists a continuous function V defined in a neighborhood of 0 and such that, for some positive constants $C_1, C_2, C_3, r_1, r_2, r_3$ and δ, with $r_3 > r_2$,*

$$\|x\| < \delta \quad \Rightarrow \quad C_1 \|x\|^{r_1} \leq V(x) \leq C_2 \|x\|^{r_2}, \tag{5.16}$$

$$\|x\| < \delta \quad \Rightarrow \quad \dot{V}(x) \leq -C_3 \|x\|^{r_3}. \tag{5.17}$$

Proof. (Sufficient part) Assume the existence of a function V as in the statement. Pick any trajectory $x(\cdot)$ of (5.1) and set again $v(t) = V(x(t))$. Then, by (5.16) and (5.17), for any $t \geq 0$

$$\overline{D^+} v(t) \leq -C_3 \left(\frac{v(t)}{C_2}\right)^{\frac{r_3}{r_2}}. \tag{5.18}$$

Integrating (5.18) and using (5.16), we obtain at once (5.15), with $k = r_1\left(\frac{r_3}{r_2} - 1\right)$, $\eta = \frac{r_2}{r_1}$, and for M some positive constant depending on C_1, C_2, C_3, r_1, r_2 and r_3.

(Necessary part) Assume that the estimate (5.15) is valid. Let $\varphi(t, x)$ denote the flow of (5.1). Pick some $r > \frac{k}{\eta}$ and set

$$V(x) := \int_0^{+\infty} \|\varphi(s, x)\|^r \, ds. \tag{5.19}$$

We first check that V is well-defined and continuous on a neighborhood of 0. It follows from (5.15) that

$$\begin{aligned}
\int_0^{+\infty} \|\varphi(s, x)\|^r \, ds &\leq M^r \|x\|^{\eta r} \int_0^{+\infty} \big(1 + \|x\|^k s\big)^{-\frac{r}{k}} \, ds \\
&= M^r \|x\|^{\eta r - k} \int_0^{+\infty} (1 + \sigma)^{-\frac{r}{k}} \, d\sigma.
\end{aligned} \tag{5.20}$$

The last integral term in (5.20) is finite since $r > k$, hence V is well-defined and the right inequality in (5.16) holds true with $r_2 = \eta r - k$, $C_2 = M^r \int_0^{+\infty} (1+\sigma)^{-\frac{r}{k}} d\sigma$. Using the continuity of the flow and applying Lebesgue's theorem, we see that V is continuous, as well. To prove the left inequality in (5.16), we fix some $x \neq 0$ and we set $h(s) = \|\varphi(s,x)\|^r$ for $s \geq 0$. Let $L > 0$ be a Lipschitz constant for f in a neighborhood of 0, so that $\|f(\varphi(s,x))\| \leq L\|\varphi(s,x)\|$, $\forall s \geq 0$. Clearly, $h'(s) = r\|\varphi(s,x)\|^{r-2} \langle \varphi(s,x), f(\varphi(s,x)) \rangle$, hence $|h'(s)| \leq L r h(s)$. We infer that

$$V(x) \geq (Lr)^{-1} \int_0^{+\infty} |h'(s)|\, ds \geq (Lr)^{-1} \left| \int_0^{+\infty} h'(s)\, ds \right| = (Lr)^{-1} \|x\|^r.$$

Hence the left inequality in (5.16) is fulfilled with $C_1 = (Lr)^{-1}$, $r_1 = r$. On the other hand, for any x,

$$\dot{V}(x) = \limsup_{t \to 0^+} \left(-\frac{1}{t} \int_0^t \|\varphi(s,x)\|^r\, ds \right) = -\|x\|^r.$$

Hence (5.17) is fulfilled, with $C_3 = 1$, $r_3 = r = r_1$. ∎

Corollary 5.1 *Let f be a vector field of class C^1 near 0, and such that $f(0) = 0$. Assume that (5.15) is fulfilled, and that for some constants $C, p, \delta > 0$,*

$$\left\|\frac{\partial \varphi}{\partial x}(t,x)\right\| \leq C(1 + \|x\|^k t)^p, \quad \forall t \geq 0,\ \|x\| < \delta. \tag{5.21}$$

($\|\cdot\|$ denotes either the usual Euclidean norm in $\mathbb{R}^n$, or the subordinate matrix norm.) Assume that $\|g(x)\| = o(\|x\|^{k+\eta+r(1-\eta)})$ as $x \to 0$. Then the origin is still AS for the perturbed system

$$\dot{x} = f(x) + g(x). \tag{5.22}$$

Proof. Let V be defined by (5.19). Then V is of class C^1, and for any x close to the origin, if r is large enough,

$$\nabla V(x)^T = \int_0^{+\infty} r\|\varphi(s,x)\|^{r-2} \varphi(s,x)^T \frac{\partial \varphi}{\partial x}(s,x)\, ds,$$

so that

$$\begin{aligned}\|\nabla V(x)\| &\leq rM^{r-1}C\|x\|^{\eta(r-1)} \int_0^{+\infty} (1 + \|x\|^k s)^{p-\frac{r-1}{k}} ds \\ &= \left(rM^{r-1}C \int_0^{+\infty} (1+\sigma)^{p-\frac{r-1}{k}} d\sigma\right) \|x\|^{\eta(r-1)-k},\end{aligned}$$

and

$$|\langle \nabla V(x), g(x) \rangle| = o(\|x\|^r).$$

Since, by (5.17) (with $r_3 = r$), $\langle \nabla V(x), f(x) \rangle \leq -C_3 \|x\|^r$, we get

$$\langle \nabla V(x), f(x) + g(x) \rangle \leq -C_3' \|x\|^r$$

for some constant $C_3' > 0$ and for all x close to 0. ∎

Remark 5.2 Even if f is a polynomial vector field, the asymptotic stability of (5.1) at the origin does not guarantee that $\|\frac{\partial \varphi}{\partial x}(t,x)\|$ is bounded for $t \geq 0$ and x close to 0.

Example 5.2 (Continued) Pick any $\varepsilon \in (0, 1/2)$, and consider the solution of (5.9) issuing from $(x_1, x_2)^T = (-\varepsilon, 0)^T$ at $t = 0$. It follows from (5.10) that

$$\frac{dr}{d\theta} = \frac{r^{\frac{5}{2}} \sin^{-\frac{1}{2}} \theta \cos^4 \theta}{1 - r^{\frac{3}{2}} \cos^3 \theta \sin^{\frac{1}{2}} \theta}$$

as long as $0 < \theta(t) < \pi$. We infer that, for some constant $C > 0$,

$$[\ln(r(\theta))]_0^\pi = \int_0^\pi r^{-1} \frac{dr}{d\theta} d\theta = \int_0^\pi \frac{r^{\frac{3}{2}} \sin^{-\frac{1}{2}} \theta \cos^4 \theta}{1 - r^{\frac{3}{2}} \cos^3 \theta \sin^{\frac{1}{2}} \theta} d\theta \leq C,$$

hence $r_{|\theta=\pi/2} \geq r_{|\theta=0} \geq e^{-C} r_{|\theta=\pi} = e^{-C}\varepsilon/\sqrt{2}$. Let t_1 be the first time for which x_1 vanishes. An application of the mean-value theorem yields

$$\sup_{-\varepsilon \leq x_1 \leq 0} \|\frac{\partial \varphi}{\partial x}(t_1, x_1, 0)\| \geq \frac{|x_2(t_1) - 0|}{|x_1(0) - 0|} = \frac{\sqrt{2 r(t_1)}}{\varepsilon} \geq 2^{\frac{1}{4}} e^{-\frac{C}{2}} \varepsilon^{-\frac{1}{2}}.$$

Therefore, for any $\varepsilon_0 \in (0, 1/2)$,

$$\sup_{t \geq 0, \|x\| \leq \varepsilon_0} \|\frac{\partial \varphi}{\partial x}(t, x)\| = +\infty.$$

∎

Next proposition gives sufficient conditions for (5.21) to hold.

Proposition 5.1 *The growth condition (5.21) is fulfilled in each of the following situations:*
(i) $f(x) = Ax$ *with A Hurwitz;*
(ii) f *is of class C^2, and $f(tx) = t^{k+1} f(x)$ for all $t > 0$, $x \in \mathbb{R}^n$.*

Indeed, if (i) is satisfied, then $\|\frac{\partial \varphi}{\partial x}(t,x)\| = \|e^{tA}\| < Const$. The proof of (5.21) when f fulfills (ii) may be found in [163, Thm. 38]. Notice, however, that for homogeneous systems the result in Corollary 5.1 may be proved in another way. (See below Subsection 5.3.2.)

5.1.3 Finite-time stability

Due to the uniqueness of solutions in backward time, the trajectories of (5.1) cannot reach the origin in *finite time* if the field f is of class C^1 (or merely, locally Lipschitz continuous). However, the application of a continuous feedback law $u = u(x)$ in a *smooth* control system

$$\dot{x} = g(x, u) \tag{5.23}$$

may result in a closed loop system

$$\dot{x} = g(x, u(x)) =: f(x) \tag{5.24}$$

for which the origin is locally asymptotically stable and it is reached in finite time by any trajectory. A trivial one-dimensional example is provided by $g(x, u) = x - u^3$, $(x, u \in \mathbb{R})$ and $u(x) = x^{\frac{1}{5}}$. (Notice that the feedback law $u(x) = (2x)^{\frac{1}{3}}$ stabilizes (5.23) exponentially, but not in finite time.)

Throughout this subsection we assume that (i) f is a continuous vector field defined on a neighborhood of 0, (ii) $f(0) = 0$, and (iii) (5.1) possesses unique solutions in forward time; that is, if two solutions agree at some time t_0, then they agree at any time $t \geq t_0$. Let $\varphi(t, x)$ denote the flow map, which is continuously defined (thanks to (iii)) on an open set in $\mathbb{R}^+ \times \mathbb{R}^n$. The following definition has been first introduced in [19].

Definition 5.3 *The origin is* finite-time stable *for (5.1) if it is stable and there exist an open neighborhood U of the origin, and a function $T : U \setminus \{0\} \to (0, +\infty)$ (called the* settling-time function*) such that, for each $x \in U \setminus \{0\}$, $\varphi(\cdot, x)$ is defined on $[0, T(x))$, $\varphi(t, x) \in U \setminus \{0\}$ $\forall t \in [0, T(x))$, and $\lim_{t \to T(x)} \varphi(t, x) = 0$.*

We set $T(0) = 0$. We stress that the settling-time function may fail to be continuous (or even bounded) at the origin. (See [19, Example 2.2].) A useful characterization of finite-time stability (together with the continuity of the settling-time function) is given in the next theorem [19].

Theorem 5.4 *Let f be as above. Then the origin is finite-time stable and the settling-time function is continuous at 0 if, and only if, there exist real numbers $C > 0$ and $\alpha \in (0, 1)$, and a continuous positive definite function V defined on an open neighborhood Ω of 0, such that*

$$\forall x \in \Omega \setminus \{0\}, \quad \dot{V}(x) \leq -C V(x)^\alpha. \tag{5.25}$$

If this is the case, then the settling-time function $T(x)$ is actually continuous in a neighborhood of 0, and it fulfills (for $\|x\|$ small enough)

$$T(x) \leq \frac{1}{C(1-\alpha)} V(x)^{1-\alpha}. \tag{5.26}$$

Sufficient conditions for the existence of a Liapunov function fulfilling (5.25) will be given in Subsection 5.3.2.

5.2 Analyticity

Throughout this section we are given an *analytic* vector field f on a neighborhood of the origin, such that $f(0) = 0$.

5.2.1 Analytic unsolvability of the stability problem

Let g be another analytic vector field defined in a neighborhood of 0 (in $\mathbb{R}^n$), and let $k \in \mathbb{N}^*$. We say that f has *kth order contact* with g at 0 (and we write $f \sim_k g$ at 0) if the Taylor expansions of f and g up to order k are identical at 0; that is

$$\frac{\partial^{|\alpha|} f_i}{\partial x^\alpha}(0) = \frac{\partial^{|\alpha|} g_i}{\partial x^\alpha}(0), \quad \forall |\alpha| \leq k, \ 1 \leq i \leq n.$$

The equivalence classes (under $\sim_k$ at 0) are called $k-jets$ (of analytic vector fields vanishing at the origin). Let j be a k–jet. If for all f in this class the origin is an AS (respectively, unstable) equilibrium position, then the jet itself is called *asymptotically stable (AS)* (respectively, *unstable*). A jet which is neither stable nor unstable is called *neutral*. We denote by $J^+_{k,n}$, $J^-_{k,n}$ and $J^0_{k,n}$ the spaces of k–jets in $\mathbb{R}^n$ which are AS, unstable or neutral, respectively. They can be reviewed as subspaces of some Euclidean space $\mathbb{R}^m$.

Example 5.3 Let j_1, j_2 and j_3 be the respective 1–jets of $x_2 \frac{\partial}{\partial x_1} - x_1 \frac{\partial}{\partial x_2}$, $-x_1 \frac{\partial}{\partial x_1} - x_2 \frac{\partial}{\partial x_2}$, and $x_1 \frac{\partial}{\partial x_1} + x_2 \frac{\partial}{\partial x_2}$. Then $j_1 \in J^0_{1,2}$, $j_2 \in J^+_{1,2}$ and $j_3 \in J^-_{1,2}$.

Example 5.2 (Continued) Let j denote the 3–jet of $(-x_1^3 + x_2^3)\frac{\partial}{\partial x_1} - x_1 \frac{\partial}{\partial x_2}$. Then $j \in J^0_{3,2}$.

One may wonder if there exists an algebraic criterion that allows us to distinguish the AS k–jets from the others. This is not the case, by virtue of the following theorem due to Arnold [4].

Theorem 5.5 *[4] Let $n \geq 3$ and $k \geq 5$. Then $J^+_{k,n}$ is not semialgebraic.*

Recall that a semialgebraic set is a set which can be defined by a finite number of polynomial equations and inequalities.

Afterwards, this result has been improved by Il'jašenko [78].

Theorem 5.6 *[78] Let $n \geq 5$ and $k \geq 5$. Then $J_{k,n}^+$ is not semianalytic.*

(A set is *semianalytic* if it may be defined, in a neighborhood of each of its points, by a finite number of analytic equations and inequalities.) Thus, in contrast to what happens in the one-dimensional case, the asymptotic stability problem fails to be analytically solvable in higher dimension.

5.2.2 Analytic Liapunov functions

Let f be an analytic vector field such that the origin is an AS equilibrium point for (5.1). A strict Liapunov function of class C^∞ exists, according to Theorem 2.4. Unfortunately, its definition involves trajectories of (5.1), which are not analytically known, in general. We may wonder if there exists a "simpler" Liapunov function, as e.g. a polynomial function or an analytic one. The following result shows that an *analytic* strict Liapunov function may fail to exist.

Proposition 5.2 *a) There exists an algebraic vector field f_0 on $\mathbb{R}^2$ such that (i) 0 is a center for (5.1), and (ii) a (non-trivial) analytic first integral fails to exist. In particular, there does not exist any analytic weak Liapunov function. b) There exists an algebraic vector field f_1 on $\mathbb{R}^2$ such that (i) 0 is an AS equilibrium point for (5.1), and (ii) an analytic strict Liapunov function fails to exist.*

Proof. a) Let $p(x_1, x_2) := 2x_1^2 + x_2^2$, $q(x_1, x_2) := x_1^2 + x_2^2$ for all $x = (x_1, x_2)^T \in \mathbb{R}^2$, and pick any $\lambda \in (0, +\infty) \setminus \mathbb{Q}$. Set $V(x_1, x_2) := p(x_1, x_2)^\lambda q(x_1, x_2)$. Then V is a positive definite function of class C^1. On the other hand,

$$dV = p^{\lambda-1}(\lambda q\, dp + p\, dq) =: p^{\lambda-1}(a\, dx_1 + b\, dx_2),$$

where $a(x_1, x_2) := 4\lambda x_1(x_1^2 + x_2^2) + 2x_1(2x_1^2 + x_2^2)$, $b(x_1, x_2) := 2\lambda x_2(x_1^2 + x_2^2) + 2x_2(2x_1^2 + x_2^2)$. (Notice that a and b are homogeneous polynomial functions of degree 3.) We take $f_0 := -b\frac{\partial}{\partial x_1} + a\frac{\partial}{\partial x_2}$, so that $L_{f_0}V = dV(f_0) = 0$. Clearly, 0 is the only equilibrium point of f_0 and it is a center. To prove (ii), we argue by contradiction. Assume that $I = \sum_{k \geq 0} I_k$ ($\neq Const.$) is an analytic first integral (decomposed as a series of homogeneous polynomial functions,

I_k being of degree k for each k). Then $0 = L_{f_0}I = \sum_{k\geq 0} L_{f_0}I_k$. Since for each k $L_{f_0}I_k$ is a homogeneous polynomial function of degree $k+2$, we infer that $L_{f_0}I_k = 0$ for each $k \geq 0$, so that we may in fact assume that I is a homogeneous polynomial function of degree $k \geq 1$. Since $L_{f_0}I = 0$, $I = Const.$ on the set $\{p^\lambda q = 1\}$. A homogeneity argument shows that, for some constant $C \neq 0$,

$$I(x_1, x_2) = C\left[(p^\lambda q)(x_1, x_2)\right]^{\frac{k}{2(\lambda+1)}} \quad \forall x = (x_1, x_2)^T \in \mathbb{R}^2$$

hence,

$$I(1, z) = C\left(2 + z^2\right)^{\frac{k\lambda}{2(\lambda+1)}} \left(1 + z^2\right)^{\frac{k}{2(\lambda+1)}} \quad \forall z \in \mathbb{R}. \tag{5.27}$$

The function $z \mapsto (2 + z^2)^{\frac{k\lambda}{2(\lambda+1)}}$ may be prolonged as a holomorphic function on $\mathbb{C} \setminus \{z = is : s \in (-\infty, -\sqrt{2}] \cup [\sqrt{2}, +\infty)\}$ ($\ni i$), and by (5.27) the same is true for the function $z \mapsto (1 + z^2)^{\frac{k}{2(\lambda+1)}}$. As a consequence, there exist an integer $n \geq 1$ and a number $\alpha \in \mathbb{C} \setminus \{0\}$ such that $(1 + z^2)^{\frac{k}{2(\lambda+1)}} \sim \alpha(z-i)^n$ as $z \to i$, which yields (when restricting to $z = is$, $s \to 1^-$) $\frac{k}{2(\lambda+1)} = n$. This contradicts the assumption $\lambda \notin \mathbb{Q}$. To complete the proof of (ii), we observe that each orbit of (5.1) is periodic, hence any weak Liapunov function turns out to be a first integral.

b) Let p, q, λ, V, a, b and f_0 be as above. We take $f_1 = f_0 - (x_1^2 + x_2^2)\left(a\frac{\partial}{\partial x_1} + b\frac{\partial}{\partial x_2}\right)$. (i) is fulfilled, since $L_{f_1}V = -(x_1^2 + x_2^2)p^{\lambda-1}(a^2 + b^2)$ is negative definite. To check (ii), assume the existence of an analytic strict Liapunov function $W = \sum_{k\geq 0} W_k$. (W_k is a homogeneous polynomial function of degree k for each k.) Let W_{k_0} be the first not vanishing term ($k_0 \geq 2$). Clearly, $L_{f_1}W = L_{f_0}W_{k_0} + R$, with the order of R larger that the one of $L_{f_0}W_{k_0}$ (namely, $k_0 + 2$). It follows that $L_{f_0}W_{k_0} \leq 0$ everywhere, hence f_0 possesses an analytic first integral, which is impossible according to a). ∎

Several interesting properties may be proved when an analytic strict Liapunov function exists. They rest upon the following well-known result [96].

Lemma 5.1 *Let $G : \mathbb{R}^n \to \mathbb{R}$ be an analytic function defined on an open neighborhood U of 0. We assume that G is positive definite, i.e., $G(0) = 0$ and $G(x) > 0$ $\forall x \in U \setminus \{0\}$. Then G satisfies a Lojasewicz inequality; that is, there exist an open set U' (with $\{0\} \subset U' \subset U$), an integer $m > 0$, and a real number $C > 0$ such that*

$$G(x) \geq C \|x\|^m, \quad \forall x \in U'. \tag{5.28}$$

The part (ii) of the following result comes from [14].

Proposition 5.3 *Let f be an analytic vector field defined on an open neighborhood Ω of 0 in $\mathbb{R}^n$ and such that $f(0) = 0$. Assume that there exists an analytic strict Liapunov function V for (5.1). Then the following properties hold true.*
(i) The origin is rationally (possibly exponentially) stable;
(ii) There exists some integer k such that for any vector field g of class C^1 fulfilling $\|g(x)\| = o(\|x\|^k)$ as $x \to 0$, the origin is still AS for (5.22);
(iii) There exists also a polynomial strict Liapunov function for (5.1).

Proof. Since $-\langle \nabla V(x), f(x) \rangle$ and $V(x)$ are analytic functions (in a neighborhood of 0) which are *positive definite*, there exist (by Lemma 5.1) an open set $\Omega' \subset \Omega$, two integers $m_1, m_2 > 0$ and two real numbers $C_1, C_2 > 0$ such that

$$-\langle \nabla V(x), f(x) \rangle \geq C_1 \|x\|^{m_1} \quad \forall x \in \Omega', \tag{5.29}$$

$$V(x) \geq C_2 \|x\|^{m_2} \quad \forall x \in \Omega'. \tag{5.30}$$

(i) Obviously, $\nabla V(0) = 0$, hence for some constant $C_3 > 0$,

$$V(x) \leq C_3 \|x\|^2 \quad \forall x \in \Omega'. \tag{5.31}$$

Applying Theorems 5.2 and 5.3, we deduce that the origin is exponentially stable or rationally stable for (5.1), according to $m_1 = 2$ or $m_1 > 2$.
(ii) Let $k = m_1 - 1$. Since $\|\nabla V(x)\| = O(\|x\|)$, $|\langle \nabla V(x), g(x) \rangle| = o(\|x\|^{m_1})$, hence

$$\langle \nabla V(x), f(x) + g(x) \rangle \leq -C_1 \|x\|^{m_1} + |\langle \nabla V(x), g(x) \rangle| \leq C_1' \|x\|^{m_1}$$

for some constant $C_1' > 0$ and for all x close enough to the origin. Therefore V is also a strict Liapunov function for $f + g$ at the origin.
(iii) Let $m = \max(m_1, m_2)$ and let V_m denote the Taylor polynomial of V of degree m. Then $|V(x) - V_m(x)| = O(\|x\|^{m+1}) = O(\|x\|^{m_2+1})$ and we infer from (5.30) that for all x in a neighborhood of 0, $V_m(x) \geq C_2' \|x\|^{m_2}$ for some constant $C_2' > 0$. On the other hand $\|\nabla V(x) - \nabla V_m(x)\| = O(\|x\|^m)$, whereas $\|f(x)\| = O(\|x\|)$, hence $|\langle \nabla V(x) - \nabla V_m(x), f(x) \rangle| = O(\|x\|^{m_1+1})$ which, combined with (5.29), yields (for some constant $C_1' > 0$ and for all x close to 0)

$$\langle \nabla V_m(x), f(x) \rangle \leq -C_1' \|x\|^{m_1}.$$

We conclude that V_m is still a strict Liapunov function for (5.1). ∎

5.2.3 Holomorphic systems

In this subsection $f : \Omega \subset \mathbb{C}^n \to \mathbb{C}^n$ is a holomorphic vector field such that $f(0) = 0$. Let $A = \left(\frac{\partial f}{\partial z}\right)_{|z=0}$ be the Jacobian matrix of f at the origin. We investigate the links between the asymptotic stability of the origin for the system

$$\frac{dz}{dt} = f(z), \qquad (5.32)$$

and the one for the linearized system

$$\frac{dz}{dt} = A\,z. \qquad (5.33)$$

(The time t is still real.) As for real systems, if (5.33) is AS at 0 (i.e., A is Hurwitz), then (5.32) is also AS at 0. Surprisingly enough, the converse is still true, as it is shown by next result due to Krikorian (which improves a previous result by Dayawansa and Martin [52]).

Theorem 5.7 *[89] Let $\Omega \subset \mathbb{C}^n$ be an open set, and let $f : \Omega \to \mathbb{C}^n$ be a holomorphic vector field such that $f(0) = 0$. If 0 is locally attractive for (5.32), then it is locally asymptotically stable for both (5.32) and (5.33).*

In other words, 0 is AS for (5.32) if, and only if, A is Hurwitz. (Recall that this condition may be checked by means of Routh's criterion.) We infer that for holomorphic systems, the asymptotic stability problem is algebraically solvable.

5.3 Weighted homogeneity

Let f be an analytic vector field on $\mathbb{R}^n$ for which the origin is an AS equilibrium position. Often, the asymptotic stability of the origin occurs for an "approximating" system, corresponding to a Taylor segment (not the homogeneous first approximation, see Example 5.1.) If the components of f are polynomial functions of degree (at most) k, then the addition of terms of degree $k+1$ in the r.h.s. of (5.1) may result in a system which is no longer AS. (See Example 5.2.) A criterion involving the rate of convergence for the trajectories of (5.1) together with the growth of the Jacobian flow (see Corollary 5.1) ensures that the origin is still AS for (5.22), provided that the perturbation g is made of high enough order terms. Here, we are interested in a robustness property which does not involve the flow of (5.1). Such a property may be obtained if the field f is homogeneous (in the generalized sense defined below). The following example demonstrates that, sometimes, it is worth giving different weights to the coordinates.

Example 5.4 Let us consider the system

$$\begin{cases} \dot{x}_1 &= x_2 - x_1^3, \\ \dot{x}_2 &= -x_1^5 + x_2^2. \end{cases} \tag{5.34}$$

For any $k \leq 5$ let f_k denote the Taylor segment of degree k. (For instance, $f_1 = x_2 \frac{\partial}{\partial x_1}$ is the linear part.) Then it is easily seen that for any $k \leq 4$ the origin fails to be AS for the approximating system

$$\dot{x} = f_k(x).$$

If, instead, the coordinates x_1 and x_2 are given different weights (namely, $r_1 = 1$ for x_1 and $r_2 = 3$ for x_2), then the homogeneous first approximation becomes

$$\begin{cases} \dot{x}_1 &= x_2 - x_1^3, \\ \dot{x}_2 &= -x_1^5. \end{cases} \tag{5.35}$$

It follows from LaSalle invariance principle (applied with $V(x) = x_1^6/6 + x_2^2/2$) that (5.35) is AS. It will follow from next development that the same is true for (5.34). ∎

Precise definitions of weights, homogeneity, etc. are given in the following subsection.

5.3.1 A few definitions

Definition 5.4 *[84] Fix a set of coordinates $(x_1, ..., x_n)$ in $\mathbb{R}^n$. Let $r = (r_1, ..., r_n)$ be a n-uplet of positive real numbers.*
* *The one-parameter family of dilations $(\delta_\varepsilon^r)_{\varepsilon > 0}$ (associated with r) is defined by*

$$\delta_\varepsilon^r(x) := (\varepsilon^{r_1} x_1, ..., \varepsilon^{r_n} x_n), \quad \forall x = (x_1, ..., x_n) \in \mathbb{R}^n, \forall \varepsilon > 0.$$

The numbers r_i are the weights *of the coordinates.*
* *A function $V : \mathbb{R}^n \to \mathbb{R}$ is said to be δ^r− homogeneous of degree m ($m \in \mathbb{R}$) if*

$$V(\delta_\varepsilon^r(x)) = \varepsilon^m V(x) \quad \forall x \in \mathbb{R}^n, \forall \varepsilon > 0.$$

* *A vector field $f = \sum_{i=1}^n f_i \frac{\partial}{\partial x_i}$ is said to be δ^r−homogeneous of degree k if the component f_i is δ^r−homogeneous of degree $k + r_i$ for each i; that is,*

$$f_i(\varepsilon^{r_1} x_1, ..., \varepsilon^{r_n} x_n) = \varepsilon^{k+r_i} f_i(x), \quad \forall x \in \mathbb{R}^n, \forall \varepsilon > 0, \forall i \in [[1, n]].$$

Example 5.4 (Continued) Pick $r = (r_1, r_2) = (1,3)$, and set $f = (x_2 - x_1^3)\frac{\partial}{\partial x_1} - x_1^5 \frac{\partial}{\partial x_2}$, and $g = x_2^2 \frac{\partial}{\partial x_2}$. Then f (respectively, g) is δ^r–homogeneous of degree 2 (respectively, 3).

If all $r_i = 1$, we write δ^1 and call this the *standard dilation*. Let us stress that, with this terminology, a homogeneous (in the usual sense) polynomial function of degree m is δ^1–homogeneous of degree m, whereas a homogeneous polynomial vector field f of degree k (i.e., $f = \sum_i f_i \frac{\partial}{\partial x_i}$ with $\deg(f_i) = k$ for each i) is δ^1– homogeneous of degree $k - 1$.

The weighted homogeneity has been first introduced by Rothschild and Stein [120] for the analysis of hypoelliptic partial differential operators. Later, this notion has been extensively used for deriving new nonlinear criteria of controllability. More precisely, for affine control systems fulfilling a geometric condition involving Lie brackets (e.g., Hermes' condition or Sussmann's condition), the construction of a nilpotent approximating system which is STLC and homogeneous with respect to a certain family of dilations (not the standard one, in general), plays a great role in the analysis. (See [74].) Here, however, we shall only be interested in the stability (or stabilization) problem.

Definition 5.5 *The (generalized) Euler vector field e associated with the family of dilations $(\delta_\varepsilon^r)_{\varepsilon>0}$ is defined by*

$$e = \sum_{i=1}^{n} r_i x_i \frac{\partial}{\partial x_i}.$$

The following infinitesimal characterization of δ^r–homogeneity proves to be useful in certain circumstances.

Proposition 5.4 *Let $(\delta_\varepsilon^r)_{\varepsilon>0}$ and e be as in Definition 5.5. Let V (respectively, f) be a function (respectively, a vector field) of class C^1 in $\mathbb{R}^n$, and let $m, k \in \mathbb{R}$. Then*
(i) *V is δ^r–homogeneous of degree m if, and only if, $e \cdot V = mV$;*
(ii) *f is δ^r–homogeneous of degree k if, and only if, $[e, f] = kf$. (Here, $[e, f] = \frac{\partial f}{\partial x} e - \frac{\partial e}{\partial x} f$.)*

The proof is classical.

Corollary 5.2 *Let $(\delta_\varepsilon^r)_{\varepsilon>0}$ be any family of dilations in $\mathbb{R}^n$. Then any analytic vector field on $\mathbb{R}^n$ can be expanded as a series of δ^r– homogeneous vector fields.*

Indeed, each monomial vector field $h = Cx^\alpha \frac{\partial}{\partial x_i}$ (with $C \in \mathbb{R}$, $\alpha \in \mathbb{N}^n$, $i \in [[1, n]]$) is δ^r-homogeneous of degree $\langle r, \alpha \rangle - r_i$, since $[e, h] = \left(\sum_{j=1}^n r_j \alpha_j - r_i\right)h$. ∎

Corollary 5.3 *Let $(\delta_\varepsilon^r)_{\varepsilon > 0}$ be any family of dilations on $\mathbb{R}^n$, and let V_1, V_2 (respectively, f_1, f_2) be δ^r-homogeneous functions (respectively, vector fields) of degrees m_1, m_2 (respectively, k_1, k_2). Then $V_1 V_2$ (respectively, $V_1 f_1$, $[f_1, f_2]$) is δ^r-homogeneous of degree $m_1 + m_2$ (respectively, $m_1 + k_1$, $k_1 + k_2$).*

Proof. The homogeneity of $V_1 V_2$, $V_1 f_1$ follows at once from Definition 5.4. For $[f_1, f_2]$, we invoke Jacobi identity and apply Proposition 5.4 (twice):

$$\begin{aligned}[e, [f_1, f_2]] &= -[f_1, [f_2, e]] - [f_2, [e, f_1]] \\ &= [f_1, k_2 f_2] + [k_1 f_1, f_2] \\ &= (k_1 + k_2)[f_1, f_2].\end{aligned}$$

∎

By definition, any norm in $\mathbb{R}^n$ is δ^1-homogeneous of degree 1. When working with δ^r-homogeneity, we are led to consider δ^r-homogeneous norms, defined as follows.

Definition 5.6 *A δ^r-homogeneous norm is a map $x \mapsto \|x\|_{r,p}$, where for any $p \geq 1$*

$$\|x\|_{r,p} := \left(\sum_{i=1}^n |x_i|^{\frac{p}{r_i}}\right)^{\frac{1}{p}} \quad \forall x \in \mathbb{R}^n.$$

The set $S_{r,p} = \{x : \|x\|_{r,p} = 1\}$ is the corresponding δ^r-homogeneous unit sphere.

Obviously, each δ^r-homogeneous norm is δ^r-homogeneous of degree 1, and positive definite. (It is not a norm in the usual sense, since it is not δ^1-homogeneous.) It may be seen that all δ^r-homogeneous norms are equivalent. Assume in addition that $r_i \in \mathbb{N}^*$ for each i. Choosing p in such a way that all r_i divide $p/2$, we get an homogeneous norm which is analytic on $\mathbb{R}^n \setminus \{0\}$. For such a choice of p, $S_{r,p}$ is an analytic $(n-1)$ dimensional submanifold in $\mathbb{R}^n \setminus \{0\}$. (For $r = (1, ..., 1)$ and $p = 2$, $S_{n,p} = S^{n-1}$.)

5.3.2 Homogeneous Liapunov functions

It is well-known that an AS linear system possesses a strict Liapunov function which is a quadratic form (hence, a δ^1-homogeneous function of degree 2.)

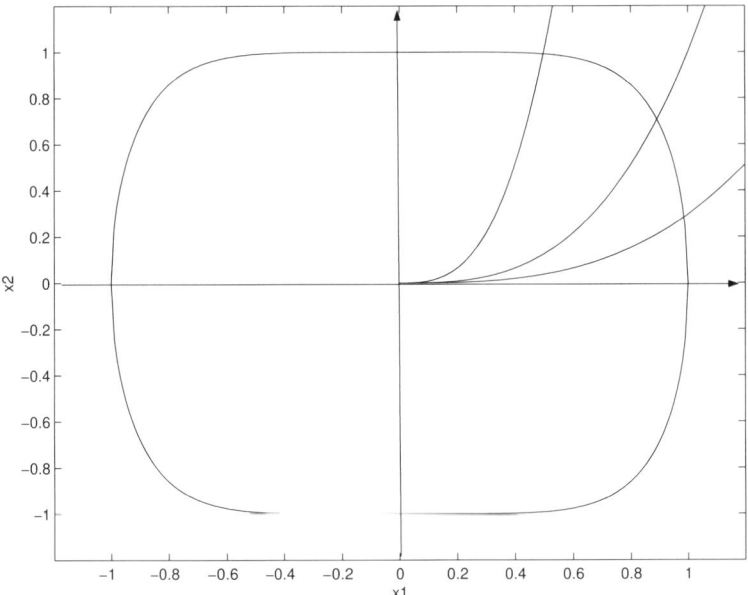

Figure 5.1: The δ^r-homogeneous unit sphere (with $n = 2$, $r = (1,3)$, $p = 6$) and some homogeneous rays

It turns out that any homogeneous AS system admits a homogeneous strict Liapunov function. A constructive proof of this result may be found in [66], [163], when the vector field f is of class C^1. When f is merely continuous, the result has been obtained by Kawski [84] when $n = 2$, and by Hermes [73] when solutions are unique in forward time. These assumptions are no longer required in the next result.

Theorem 5.8 *[115] Let f be a continuous vector field on $\mathbb{R}^n$ such that the origin is a locally AS equilibrium point. Assume that f is δ^r–homogeneous of degree k for some $r \in (0, +\infty)^n$. Then, for any $p \in \mathbb{N}^*$ and any $m > p \cdot \max_i \{r_i\}$, there exists a strict Liapunov function V for (5.1), which is δ^r–homogeneous of degree m and of class C^p. As a direct consequence, the time-derivative $\dot{V} = \langle \nabla V, f \rangle$ is δ^r–homogeneous of degree $m + k$.*

The reader is referred to Subsection 5.4.3 for the proof of a more general result (namely, Theorem 5.13).

The following corollary shows that the rate of convergence of trajectories for a homogeneous AS system is completely characterized by the degree of the field. (The last part comes from [75].)

Corollary 5.4 Let f be as in Theorem 5.8, and let $\|\cdot\|_{r,p}$ be any δ^r-homogeneous norm.

- If $k > 0$ (k is the degree of f), then there exist constants $M_1, M_2 > 0$ such that, for any trajectory $x(\cdot)$ of (5.1), for all $t \geq 0$,

$$M_1\left(1 + \|x(0)\|_{r,p}^k t\right)^{-\frac{1}{k}} \|x(0)\|_{r,p} \leq \|x(t)\|_{r,p} \leq M_2\left(1 + \|x(0)\|_{r,p}^k t\right)^{-\frac{1}{k}} \|x(0)\|_{r,p}. \tag{5.36}$$

- If $k = 0$, then there exist constants $M_1, M_2, D > 0$ such that

$$M_1 \exp(-Dt)\|x(0)\|_{r,p} \leq \|x(t)\|_{r,p} \leq M_2 \exp(-Dt)\|x(0)\|_{r,p}, \quad \forall t \geq 0. \tag{5.37}$$

- If $k < 0$, then the origin is finite-time stable.

Proof. Pick some number $m > \max\{r_1, ..., r_n, -k\}$. According to Theorem 5.8, there exists a strict Liapunov function V for (5.1), which is δ^r-homogeneous of degree m and of class C^1. Also, $\dot{V}$ is δ^r-homogeneous of degree $m+k$. V being positive definite and continuous, there exist constants $C_1, C_2 > 0$ such that

$$C_1 \leq V(x) \leq C_2, \quad \forall x \in S_{r,p}.$$

Since any $x \neq 0$ may be written as $x = \delta^r_\varepsilon(\delta^r_{\varepsilon^{-1}} x)$, with $\delta^r_{\varepsilon^{-1}} x \in S_{r,p}$ for the choice $\varepsilon := \|x\|_{r,p}$, we infer from the homogeneity of V that

$$C_1 \|x\|_{r,p}^m \leq V(x) \leq C_2 \|x\|_{r,p}^m \quad \forall x. \tag{5.38}$$

The same argument shows that for some constants $C_3, C_4 > 0$,

$$C_3 \|x\|_{r,p}^{m+k} \leq -\dot{V}(x) \leq C_4 \|x\|_{r,p}^{m+k} \quad \forall x. \tag{5.39}$$

It follows that for some constants $C_5, C_6 > 0$

$$C_5 V(x)^{\frac{m+k}{m}} \leq -\dot{V}(x) \leq C_6 V(x)^{\frac{m+k}{m}} \quad \forall x. \tag{5.40}$$

If $k > 0$ (respectively, $k = 0$), a direct integration of (5.40) combined with (5.38) gives at once (5.36) (respectively, (5.37)). The result for $k < 0$ follows from (5.40) and Theorem 5.4. ∎

Next result is often referred to as Hermes' theorem. It has been proved in [73] when solutions are unique in forward time, and in [115] without this assumption.

Corollary 5.5 Let f be as in Theorem 5.8, and let $g = \sum_{i=1}^n g_i \frac{\partial}{\partial x_i}$ be another continuous vector field on $\mathbb{R}^n$. Assume that

$$\forall i, \quad g_i(\delta^r_\varepsilon x) = o(\varepsilon^{k+r_i}) \quad \text{as } \varepsilon \to 0 \tag{5.41}$$

uniformly with respect to x on S^{n-1}. Then the origin is still AS for (5.22). Furthermore, the conclusions drawn in Corollary 5.4 are still valid for the trajectories of (5.22).

Proof. Pick a δ^r-homogeneous Liapunov function V of degree $m > \max_i r_i$ and of class C^1, as given in Theorem 5.8. Then, for $x \in S^{n-1}$, $\langle \nabla V(\delta_\varepsilon^r x), g(\delta_\varepsilon^r x) \rangle = o(\varepsilon^{m+k})$ as $\varepsilon \to 0$, whereas $\langle \nabla V(\delta_\varepsilon^r x), f(\delta_\varepsilon^r x) \rangle = \varepsilon^{m+k} \langle \nabla V(x), f(x) \rangle$. It follows that for $\varepsilon > 0$ small enough (say, $0 < \varepsilon < \varepsilon_0$) and for all $x \in S^{n-1}$,

$$\frac{3}{2}\varepsilon^{m+k}\langle \nabla V(x), f(x)\rangle < \langle \nabla V(\delta_\varepsilon^r x), (f+g)(\delta_\varepsilon^r x)\rangle < \frac{1}{2}\varepsilon^{m+k}\langle \nabla V(x), f(x)\rangle.$$

In other words, if $\dot{V}$ denotes here the time derivative of V along (5.22), if we set $\Omega := \{0\} \cup \{\delta_\varepsilon^r x : 0 < \varepsilon < \varepsilon_0, x \in S^{n-1}\}$ (Ω is a neighborhood of 0), then (5.39) and (5.40) hold true for all $x \in \Omega$. This completes the proof. ∎

5.3.3 Application to the stabilization problem

Homogeneous feedback law

For the sake of simplicity, consider a single-input, affine control system in $\mathbb{R}^n$

$$\dot{x} = f_0(x) + u\, f_1(x), \tag{5.42}$$

with f_0, f_1 real analytic vector fields such that $f_0(0) = 0$ and $f_1(0) \neq 0$. Let a family of dilations $(\delta_\varepsilon^r)_{\varepsilon > 0}$ be given, and expand f_0 and f_1 into series of δ^r-homogeneous vector fields, say $f_0 = \sum_{k \geq k_0} f_0^k$ and $f_1 = \sum_{k \geq k_1} f_1^k$. Consider the approximating system

$$\dot{x} = f_0^{k_0}(x) + u\, f_1^{k_1}(x). \tag{5.43}$$

If (5.43) is a "good" approximation of (5.42) (e.g., if (5.43) inherits the small-time local controllability of (5.42)), then it is natural to try to stabilize (5.43) by means of a δ^r-homogeneous feedback law u of degree $k_0 - k_1$. If this can be done, then it follows from Corollary 5.5 that u stabilizes also (5.42). This strategy, which can be seen as a nonlinear generalization of the classical linearization procedure, has been used in many papers. (See [85], [86], [73], [74] to mention but a few.) In particular, Kawski [85] has proved that any STLC *planar* system (5.42) admits a homogeneous approximation (5.43) which is stabilizable by means of a Hölder continuous (not C^1) homogeneous feedback law. (See Theorem 2.26, Section 2.5.)

As stabilizable linear systems may always be stabilized by means of linear feedback laws, one may wonder if a homogeneous stabilizable system may be stabilized by means of some *homogeneous* feedback law; that is, if $f: \mathbb{R}^{n+1} \to \mathbb{R}^n$ is a continuous map such that, for some numbers $r_1, ..., r_{n+1} > 0$ and some $k \in \mathbb{R}$,

$$f_i(\varepsilon^{r_1} x_1, ..., \varepsilon^{r_n} x_n, \varepsilon^{r_{n+1}} u) = \varepsilon^{k+r_i} f_i(x_1, ..., x_n, u)$$
$$\forall x = (x_1, ..., x_n) \in \mathbb{R}^n, \ \forall u \in \mathbb{R}, \ \forall \varepsilon > 0, \ \forall i \in [[1, n]],$$

and such that the system

$$\dot{x} = f(x, u) \tag{5.44}$$

is (locally) stabilized around 0 by means of some continuous feedback law $u : \mathbb{R}^n \to \mathbb{R}$, then the feedback may also be chosen to satisfy

$$u(\varepsilon^{r_1} x_1, ..., \varepsilon^{r_n} x_n) = \varepsilon^{r_{n+1}} u(x_1, ..., x_n) \quad \forall x, \ \forall \varepsilon > 0.$$

Actually, it fails to be true in general. First of all, it has been noticed that the restriction to homogeneous feedback introduces extra necessary conditions for the stabilization problem (see [86], [50], [127]). Then, a counterexample was given in [117], for $n = 2$, $r_1 = r_2 = r_3 = 1$, and f of class C^p ($p \in \mathbb{N}^*$ arbitrary). Thereafter, an algebraic counterexample has been proposed in [126]. To be precise, it is proved in [126] that the following planar system

$$\begin{cases} \dot{x}_1 &= x_1 + u, \\ \dot{x}_2 &= 3x_2 + x_1 u^2 \end{cases} \tag{5.45}$$

is STLC, stabilizable, while not stabilizable by a homogeneous feedback (i.e., a feedback u fulfilling $u(\varepsilon x_1, \varepsilon^3 x_2) = \varepsilon u(x)$). (Recall that a counterexample for which the system is planar, STLC, and *affine* in the control cannot be found, according to Kawski's theorem.) It is also proved in [126] that the following three-dimensional affine system

$$\begin{cases} \dot{x}_1 &= x_1 + x_3, \\ \dot{x}_2 &= 3x_2 + x_1 x_3^2, \\ \dot{x}_3 &= v \end{cases} \tag{5.46}$$

is STLC, stabilizable, while not stabilizable by homogeneous feedback. Observe that (5.46) is derived from (5.45) by adding an integrator. To prove the stabilizability of (5.46), the authors show that the feedback u which has been designed to stabilize (5.45) fulfills the assumptions of Proposition 2.19.

Cascade systems

The following nice result is due to Coron and Praly [48].

Theorem 5.9 *Consider the single input system*

$$\begin{cases} \dot{x}_1 = f_1(x_1) + x_2^{p_2} \\ \dot{x}_j = f_j(x_1, ..., x_j) + x_{j+1}^{p_{j+1}}, & 1 < j < n, \\ \dot{x}_n = f_n(x_1, ..., x_n) + u^{p_{n+1}}, \end{cases} \quad (5.47)$$

where the p_j's are odd integers, $f_j \in C^\infty(\mathbb{R}^j, \mathbb{R})$ and $f_j(0, ..., 0) = 0$ for all $j \in [[1, n]]$. Then (5.47) is locally stabilizable by means of a continuous feedback law.

Its proof rests on the following key result, which is a "homogeneous" version of Tsinias' theorem. (We emphasize that here the feedback laws are merely continuous, but homogeneous.)

Proposition 5.5 *[48] Let $f = (f_i)_{i=1,n} : \mathbb{R}^n \times \mathbb{R} \to \mathbb{R}^n$ be a C^0 map such that, for some numbers $r_1, ..., r_{n+1} > 0$ and some $k \in (-\min_i\{r_i\}, +\infty)$,*

$$f_i(\varepsilon^{r_1} x_1, ..., \varepsilon^{r_n} x_n, \varepsilon^{r_{n+1}} u) = \varepsilon^{k+r_i} f_i(x_1, ..., x_n, u)$$
$$\forall x = (x_1, ..., x_n) \in \mathbb{R}^n, \ \forall u \in \mathbb{R}, \ \forall \varepsilon > 0, \ \forall i \in [[1, n]].$$

Assume that the system $\dot{x} = f(x, u)$ is locally stabilizable with a continuous feedback law $u : \mathbb{R}^n \to \mathbb{R}$ such that

$$u(\varepsilon^{r_1} x_1, ..., \varepsilon^{r_n} x_n) = \varepsilon^{r_{n+1}} u(x), \qquad \forall x \in \mathbb{R}^n, \ \forall \varepsilon > 0.$$

Under these conditions, the system $\dot{x} = f(x, y)$, $\dot{y} = v$ is globally stabilizable with a continuous feedback law $v : \mathbb{R}^{n+1} \to \mathbb{R}$ such that

$$v(\varepsilon^{r_1} x_1, ..., \varepsilon^{r_n} x_n, \varepsilon^{r_{n+1}} y) = \varepsilon^{k+r_{n+1}} v(x_1, ..., x_n, y)$$
$$\forall x \in \mathbb{R}^n, \ \forall y \in \mathbb{R}, \ \forall \varepsilon > 0.$$

For a proof of Proposition 5.5 based on Theorem 5.8, see [115].

Time varying feedback

Let us consider a driftless control system of the form

$$\dot{x} = u_1 f_1(x) + \cdots + u_m f_m(x), \quad (5.48)$$

where each vector field f_i is analytic in $\mathbb{R}^n$ ($m \leq n$). Assume that $\text{rank}(f_1(x), ..., f_m(x)) = m$ and $\text{rank Lie}\{f_1, ..., f_m\}(x) = n$ for all $x \in \mathbb{R}^n$.

(Here, Lie$\{f_1,...,f_m\}$ denotes the Lie algebra spanned by the vector fields $f_1,...,f_m$.) Then (5.48) is controllable, by Chow's theorem, and stabilizable by means of a smooth time varying feedback law, according to Coron's result [44]. However, application of a *smooth* time varying feedback law may result in a slow convergence. Several studies (originating with [99], and culminating with [103]) have demonstrated that the speed of convergence may be improved by using a continuous time varying feedback law which is homogeneous (with respect to x). Assume, for simplicity, that the vector fields in (5.48) are homogeneous of degree -1 with respect to a family of dilations $(\delta_\varepsilon^r)_{\varepsilon>0}$, and that there exists a stabilizing feedback law $u = u(t,x)$ which is periodic (with respect to t) and δ^r–homogeneous (with respect to x) of degree 1. Then the following estimate occurs for the trajectories of the closed-loop system

$$\|x(t)\|_{r,p} \leq M\|x(0)\|_{r,p}\, e^{-Dt}, \quad \forall t \geq 0. \tag{5.49}$$

($\|\cdot\|_{r,p}$ denotes any δ^r–homogeneous norm.) The key result to prove such an estimate is the following extension of Theorem 5.8.

Theorem 5.10 *[107] Suppose that the origin is an AS equilibrium position of*

$$\dot{x} = F(t,x),$$

with $F \in C^0(\mathbb{R}_t \times \mathbb{R}_x^n, \mathbb{R}^n)$ being T–periodic with respect to t (i.e., $F(t+T,x) = F(t,x)$ $\forall t, x$) and δ^r–homogeneous with respect to x of degree 0 for some $r \in (0,+\infty)^n$ (i.e., $F_i(t, \delta_\varepsilon^r x) = F_i(t,x)$ $\forall (t,x) \in \mathbb{R} \times \mathbb{R}^n$, $\forall \varepsilon > 0$, $\forall i$). Then, for any $p \in \mathbb{N}$ and any $m > p\max_i\{r_i\}$ there exists a function $V(t,x)$ such that
(i) $V \in C^p(\mathbb{R} \times \mathbb{R}^n) \cap C^\infty(\mathbb{R} \times (\mathbb{R}^n \setminus \{0\}))$;
(ii) $0 < V(t,x)$ $\forall t, \forall x \neq 0$;
(iii) V *is T–periodic*: $V(t+T,x) = V(t,x)$ $\forall t, \forall x$;
(iv) V *is δ^r–homogeneous of degree m*: $V(t,\delta_\varepsilon^r x) = \varepsilon^m V(t,x)$ $\forall t, \forall x, \forall \varepsilon > 0$;
(v) $\dot{V}(t,x) := \frac{\partial V}{\partial t}(t,x) + \langle \nabla V(t,x), F(t,x)\rangle < 0$ $\forall t \in \mathbb{R}, \forall x \neq 0$.

Notice that, by (ii), (iii) and (iv), there exist positive constants C_1, C_2 such that

$$C_1\|x\|_{r,p}^m \leq V(t,x) \leq C_2\|x\|_{r,p}^m \quad \forall t, x. \tag{5.50}$$

($\|\cdot\|_{r,p}$ is any δr–homogeneous norm in $\mathbb{R}^n$.) Also, there exists a positive constant $C_3 > 0$ such that

$$\dot{V}(t,x) \leq -C_3\|x\|_{r,p}^m, \quad \forall t, x. \tag{5.51}$$

Then, (5.49) follows at once from (5.50)-(5.51).

190   Additional properties

The next result may be applied to design a homogeneous feedback law, when a (non-homogeneous) feedback law is already known.

Theorem 5.11 *[100] Let $f_1,...,f_m$ be analytic vector fields on $\mathbb{R}^n$, which are δ^r−homogeneous of degree -1 for some family of dilations $(\delta^r_\varepsilon)_{\varepsilon>0}$. Assume that (5.48) is stabilized by means of smooth, T−periodic feedback laws $u_i(t,x)$, $1 \le i \le m$, and that for some constant $C > 0$ and some T−periodic strict Liapunov function $V(t,x)$ (of class C^1) for (5.48), the level sets parametrized by t*

$$G_t^C = \{x : V(t,x) = C\}$$

are transversal to the δ^r−Euler vector field. Let $\rho : \mathbb{R} \times \mathbb{R}^n \to \mathbb{R}^+$ be the (uniquely defined) δ^r−homogeneous function of degree 1 fulfilling

$$\rho(t,x) = 1, \quad \forall t, \ \forall x \in G_t^C,$$

and let $\gamma_t : \mathbb{R}^n \setminus \{0\} \to G_t^C$, for each $t \in \mathbb{R}$, be defined by

$$\{\gamma_t(x)\} = \{\delta^r_\varepsilon x;\ \varepsilon > 0\} \cap G_t^C, \quad \forall x \neq 0.$$

Then the following feedback laws

$$\tilde{u}_i(t,x) = \rho(t,x)\, u_i(t, \gamma_t(x)), \quad i = 1,...,m$$

stabilize (5.48), while being T−periodic with respect to t and δ^r- homogeneous of degree 1 with respect to x. As a consequence, (5.49) holds true for the trajectories of the closed-loop system $\dot{x} = \sum_i \tilde{u}_i(t,x) f_i(x)$.

Finite-time stabilization

Let a control system (5.42) be given. To achieve a finite-time stabilization, it is sufficient to design a stabilizing feedback law u such that the resulting closed-loop system is homogeneous of *negative* degree with respect to some family of dilations. The following example comes from [18].

Example 5.5 The double integrator

$$\begin{cases} \dot{x}_1 &= x_2, \\ \dot{x}_2 &= u \end{cases}$$

is finite-time stabilized under the following feedback law

$$u(x_1, x_2) := -\text{sign}(x_2)|x_2|^\alpha - \text{sign}\big(\phi_\alpha(x_1,x_2)\big)\, |\phi_\alpha(x_1,x_2)|^{\frac{\alpha}{2-\alpha}}$$

for every $\alpha \in (0,1)$, where $\phi_\alpha(x_1, x_2) := x_1 + \frac{1}{2-\alpha}\text{sign}(x_2)|x_2|^{2-\alpha}$. Here, the closed-loop system is δ^r–homogeneous of degree $\alpha - 1$, for the choice $(r_1, r_2) = (2-\alpha, 1)$. Notice that an explicit homogeneous Liapunov function is also given in [18]. An output feedback finite-time stabilizing control law (inspired by this feedback law) is proposed in [75].

Another (somewhat simpler) finite-time stabilizer for the double integrator is as follows.

Example 5.6 Let us consider the system

$$\begin{cases} \dot{x}_1 = x_2, \\ \dot{x}_2 = -\text{sign}(x_1)|x_1|^\alpha - \text{sign}(x_2)|x_2|^\beta \end{cases}$$

where $\alpha > 0$, $\beta > 0$. Taking $V(x) = |x_1|^{1+\alpha}/(1+\alpha) + x_2^2/2$ as a Liapunov function candidate and applying LaSalle invariance principle in its extended form [60], we see that the origin is AS for each pair (α, β). On the other hand, the vector field $x_2 \frac{\partial}{\partial x_1} - (\text{sign}(x_1)|x_1|^\alpha + \text{sign}(x_2)|x_2|^\beta)\frac{\partial}{\partial x_2}$ is homogeneous with respect to some family of dilations if, and only if, $\beta = \frac{2}{1+\alpha^{-1}}$. If this condition is fulfilled, then (up to a factor) the weights are $(r_1, r_2) = (1, \frac{\alpha+1}{2})$, and the degree is $k = \frac{\alpha-1}{2}$. It follows from Corollary 5.4 that the origin is finite-time stable whenever $\alpha \in (0,1)$ and $\beta = \frac{2}{1+\alpha^{-1}}$. (Notice that we don't need to construct an explicit homogeneous strict Liapunov function.) A quite different analysis [68] shows that the system is still finite-time stable when $\alpha \in (0,1)$ and $0 < \beta < \frac{2}{1+\alpha^{-1}}$. ■

More generally, for n-dimensional cascade systems it has been proved in the recent paper [76] that the problem of *global finite-time* stabilization is solvable by non-Lipschitz continuous feedback laws.

Theorem 5.12 *Consider the single input system*

$$\begin{cases} \dot{x}_1 = x_2 + f_1(x_1) \\ \dot{x}_j = x_{j+1} + f_j(x_1, ..., x_j), & 1 < j < n, \\ \dot{x}_n = u + f_n(x_1, ..., x_n), \end{cases} \quad (5.52)$$

where for each j $f_j : \mathbb{R}^j \to \mathbb{R}$ is a function of class C^1 with $f_j(0, ..., 0) = 0$. Then (5.52) is globally stabilizable in finite time by means of a continuous feedback law.

5.4 Symmetries and Liapunov functions

Let f be an analytic vector field on $\mathbb{R}^n$ such that the origin is an AS equilibrium position for (5.1). According to Hermes' theorem (Corollary 5.5), to prove that the origin is indeed AS for (5.1), it is sufficient to find some n–uplet $r = (r_1, ..., r_n)$ of positive numbers such that the origin is already AS in the first δ^r–homogeneous approximation. In [74], the author asks the question whether there is an algorithm for finding, for a given real analytic vector field f that has 0 as an AS equilibrium position, a family of dilations $(\delta_\varepsilon^r)_{r>0}$ and an analytic change of coordinates ϕ such that, if one expands f in the new coordinates into a series of δ^r–homogeneous terms, then the leading term in the expansion already has 0 as an AS equilibrium position. The algorithmic question is of course related to the even more basic issue whether δ_ε^r and ϕ exist at all. A counterexample was communicated to the second author by H. Sussmann in 1995, namely the field $f := -x_1^3(x_2^2 + x_1^4)\frac{\partial}{\partial x_1} - x_2^3(x_1^2 + x_2^4)\frac{\partial}{\partial x_2}$ in $\mathbb{R}^2$. Next counterexample is still simpler.

Example 5.7 Let us consider the following system (which may be seen as a stabilized double integrator)

$$\begin{cases} \dot{x}_1 &= x_2, \\ \dot{x}_2 &= -x_1 - x_1^2 x_2. \end{cases} \quad (5.53)$$

The origin is globally AS, thanks to LaSalle invariance principle. Let $r = (r_1, r_2) \in (0, +\infty)^2$ and let a (local) analytic change of coordinates $x = \phi(\xi)$ be given. Expand $f := x_2 \frac{\partial}{\partial x_1} - (x_1 + x_1^2 x_2)\frac{\partial}{\partial x_2}$ into a series of δ^r–homogeneous terms in the new coordinates $\xi = (\xi_1, \xi_2)$, and denote by $\tilde{f}$ the leading term in the expansion. Writing $x_1 = a\xi_1 + b\xi_2 + \cdots$, $x_2 = c\xi_1 + d\xi_2 + \cdots$ (where "$\cdots$" stands for higher order terms), we have $ad - bc \neq 0$ (since ϕ is a diffeomorphism), hence $ad \neq 0$ or $bc \neq 0$. Exchanging ξ_1 and ξ_2 if needed, we may assume that $ad \neq 0$. Clearly, the degree of each term in the expansion of the function $-x_1^2 x_2$ (expressed in the new coordinates) is larger than the degree of the leading term in the expansion of $-x_1$ (or of x_2). If $r_1 = r_2$, then in the (new) coordinates $\tilde{x}_1 := a\xi_1 + b\xi_2$, $\tilde{x}_2 := c\xi_1 + d\xi_2$, $\tilde{f}$ takes the form $\tilde{f} = \tilde{x}_2 \frac{\partial}{\partial \tilde{x}_1} - \tilde{x}_1 \frac{\partial}{\partial \tilde{x}_2}$. Therefore, the origin is not AS for $\tilde{f}$. If $r_1 > r_2$, then we easily see that $\tilde{f} \cdot \xi_1 = \frac{b^2+d^2}{ad-bc} \xi_2$ (since $d \neq 0$). It follows that the degree of $\tilde{f}$ has to be *negative*. By virtue of Corollary 5.4, if the origin is AS for $\tilde{f}$, then it is actually *finite-time* stable. This cannot occur, since $\tilde{f}$ is of class C^1 near 0 (hence the only trajectory reaching the origin is $\xi(t) \equiv 0$). A similar argumentation may be used when $r_1 < r_2$. ∎

In this example, no one homogeneous approximation may constitute a "good" approximation of the original system, since the asymptotic stability of the origin is not preserved. For this reason we are sometimes led to consider more general symmetries (instead of weighted homogeneity) for the system (5.1).

5.4.1 Discrete symmetry

To be short, we limit ourselves to discrete symmetries which look like rotations. It is clear that if (5.1) is AS and odd, then an *even* strict Liapunov function exists. This result may be generalized as follows.

Proposition 5.6 *Let* $f \in C^0(\mathbb{R}^n, \mathbb{R}^n)$ *and* $M \in \mathbb{R}^{n \times n}$ *be such that* $M^k = I$ *for some* $k \geq 1$, *and*

$$f(Mx) = Mf(x), \quad \forall x. \tag{5.54}$$

If 0 is globally AS for (5.1), then there exists a strict Liapunov function V *of class* C^∞ *on* $\mathbb{R}^n$, *and such that*

$$V(Mx) = V(x), \quad \forall x. \tag{5.55}$$

Proof. It is sufficient to pick any strict Liapunov function W of class C^∞ and to set $V(x) := \sum_{i=0}^{k-1} W(M^i x)$. ∎

This result may easily be extended to the case where (5.54) is fulfilled for each M in a finite subgroup of $GL_n(\mathbb{R})$.

5.4.2 Infinitesimal symmetry

The geometric theory of symmetry, which goes back to Lie, has been developed to understand and analyze the standard procedure of explicit integration of ODE. (The reader is referred to [24] for a modern exposition of the theory.) Here, to avoid useless generality, we shall consider a restricted class of symmetries, which is easily defined and contains the (previously defined) weighted homogeneity. We shall make the following standing assumption: s is a vector field of class C^1 on $\mathbb{R}^n$ which is complete, i.e., its flow (denoted by φ) is defined on $\mathbb{R}_t \times \mathbb{R}_x^n$.

Proposition 5.7 *[87] Let* $V : \mathbb{R}^n \to \mathbb{R}$ *be a function of class* C^1 *and let* $m \in \mathbb{R}$. *Then the following statements are equivalent*

(i) $s \cdot V = mV$;

(ii) $V(\varphi(\ln(\varepsilon), x)) = \varepsilon^m V(x)$, $\quad \forall \varepsilon > 0$, $\forall x \in \mathbb{R}^n$.

If any of the above is fulfilled, then s is called a *symmetry* of V, and V is said to be s−*homogeneous* of degree m. An analogous result holds true for vector fields.

Proposition 5.8 *[117], [87] Let f be a vector field of class C^1 on $\mathbb{R}^n$ and let $k \in \mathbb{R}$. Let ψ denote the flow of f. Then the following statements are equivalent*

(i) $[s, f] = kf$;

(ii) $\psi(t_2, \varphi(t_1, x)) = \varphi(t_1, \psi(e^{kt_1}t_2, x))$, whenever each term is defined.

If any of the above is fulfilled, then s is called a *symmetry* of f, and f is said to be s−*homogeneous* of degree k.

Remark 5.3 Notice that above definitions are coordinate-free.

Example 5.8 Let $r = (r_1, ..., r_n) \in (0, +\infty)^n$ be given, and assume that f is δ^r−homogeneous of degree k. Take as symmetry candidate the Euler vector field $s = \sum_{i=1}^{n} r_i x_i \frac{\partial}{\partial x_i}$. Then f is s−homogeneous of degree k, according to Proposition 5.4.

Example 5.9 Consider the system

$$\begin{cases} \dot{x}_1 = -x_1^3, \\ \dot{x}_2 = -x_2. \end{cases}$$

We claim that whatever the change of coordinates $x = \phi(\xi)$ and the family of dilations $(\delta_\varepsilon^r)_{\varepsilon > 0}$ (in the new coordinates ξ) the origin is no longer AS for the first approximating system. Indeed, writing $x_1 = a\xi_1 + b\xi_2 + \cdots$, $x_2 = c\xi_1 + d\xi_2 + \cdots$ (with $ad \neq 0$), and $f = -x_1^3 \frac{\partial}{\partial x_1} - x_2 \frac{\partial}{\partial x_2}$, we observe that the monomial ξ_2 appears in the ξ_2-component of f, so that the leading δ^r−homogeneous vector field in the expansion of f has degree 0. If the origin is AS for the approximating system, then $\xi(t)$ has to decrease exponentially. The same has to be true for $x(t)$, which is absurd.

Let $V(x) := e^{-x_1^{-2}} + x_2^2$. Then $f \cdot V = -2V$. Set $s = -f$ and $F = Vf$. Then the vector field F (of class C^∞ on $\mathbb{R}^2$) is s−homogeneous of degree 2. ∎

Caution. In contrast to what happens for weighted homogeneity, an analytic vector field may fail to admit an expansion into a series of homogeneous terms with respect to a given symmetry.

Symmetries and Liapunov functions   195

Example 5.10 Pick $s = (x_1 + \sigma x_2)\frac{\partial}{\partial x_1} + x_2 \frac{\partial}{\partial x_2}$ ($\sigma > 0$ is some constant) and assume that the vector field $g = a_1 \frac{\partial}{\partial x_1} + a_2 \frac{\partial}{\partial x_2}$ fulfills the relation $[s,g] = kg$, for some number k. It means that

$$\begin{pmatrix} \frac{\partial a_1}{\partial x_1} & \frac{\partial a_1}{\partial x_2} \\ \frac{\partial a_2}{\partial x_1} & \frac{\partial a_2}{\partial x_2} \end{pmatrix} \begin{pmatrix} x_1 + \sigma x_2 \\ x_2 \end{pmatrix} - \begin{pmatrix} 1 & \sigma \\ 0 & 1 \end{pmatrix} \begin{pmatrix} a_1 \\ a_2 \end{pmatrix} = k \begin{pmatrix} a_1 \\ a_2 \end{pmatrix}. \quad (5.56)$$

Expanding g into a series of δ^1-homogeneous terms, we see that (5.56) is also fulfilled for each term in the series; hence, we may assume that g is δ^1-homogeneous of degree k' (k' is some number). Therefore, for each $i = 1, 2$, $\sum_{j=1,2} x_j \frac{\partial a_i}{\partial x_j} = (k'+1)a_i$, and (5.56) reduces to

$$\begin{cases} x_2 \frac{\partial a_1}{\partial x_1} - a_2 = \sigma^{-1}(k - k')a_1, \\ x_2 \frac{\partial a_2}{\partial x_1} = \sigma^{-1}(k - k')a_2. \end{cases} \quad (5.57)$$

This leads to $k = k'$ and $a_2 = cx_2^{k+1}$, then $a_1 = cx_1 x_2^k + c' x_2^{k+1}$ for some constants c, c'. The only s-homogeneous vector fields of degree k take the form

$$g = cx_2^k \left(x_1 \frac{\partial}{\partial x_1} + x_2 \frac{\partial}{\partial x_2} \right) + c' x_2^{k+1} \frac{\partial}{\partial x_1}.$$

Therefore, the only analytic vector fields which can be expanded into a series of s-homogeneous terms may be written as

$$f = h_1(x_2)\left(x_1 \frac{\partial}{\partial x_1} + x_2 \frac{\partial}{\partial x_2} \right) + h_2(x_2) x_2 \frac{\partial}{\partial x_1}$$

for some analytic functions h_1 and h_2. Notice that if we perturb slightly s into $s_\varepsilon = (x_1 + \sigma x_2)\frac{\partial}{\partial x_1} + (1+\varepsilon)\frac{\partial}{\partial x_2}$ (with $0 < |\varepsilon| < 1$), then the homogeneous expansion property is recovered, for there exists another set of coordinates (y_1, y_2) in which $s_\varepsilon = y_1 \frac{\partial}{\partial y_1} + (1+\varepsilon) y_2 \frac{\partial}{\partial y_2}$. ∎

If we are given an analytic vector field f, the space

$$\mathcal{S} = \{ s \in C^\omega(\mathbb{R}^n, \mathbb{R}^n) : \exists k \in \mathbb{R}, \ [s, f] = kf \}$$

of the (analytic) symmetries of f is clearly a Lie subalgebra of the Lie algebra of analytic vector fields. Finding a (not collinear) symmetry is easy to do in one dimension (use the variation of parameter method), but highly non-trivial in the higher dimension. When (5.1) is a chain of integrators (i.e., $f_i(x) = x_{i+1}$ for $i < n$), then the generating function method [24, p. 29] transforms the search for symmetries into the integration of a n-order (variable coefficients) linear ODE. If a finite-dimensional Lie algebra L of vectors fields containing f may be found, then the existence of symmetries in L is quite easy to decide.

Proposition 5.9 *[117] Let L be a finite-dimensional Lie algebra of vector fields on $\mathbb{R}^n$, and let $f \in L \setminus (0)$. Define the operator D_f by $D_f(g) = [f, g]$, $\forall g \in L$.*
(i) If there exists $s \in L$ such that $[s, f] = f$, then D_f is nilpotent as an element of $\mathrm{End}(L)$.
(ii) Conversely, if D_f is nilpotent, then there exists $s \in L \setminus \mathrm{span}\{f\}$ such that

$$[s, f] \in \mathrm{span}\{f\}.$$

If, moreover, L is a nilpotent (respectively, semisimple) Lie algebra, then we have $[s, f] = 0$ (respectively, $[s, f] = f$).

According to [83], a nilpotent (finite-dimensional) Lie algebra of analytic vector fields is nothing else than the space spanned by the δ^r−homogeneous (polynomial) vector fields of negative degree, for a convenient choice of coordinates and of a family of dilations $(\delta^r_\varepsilon)_{\varepsilon > 0}$.

Example 5.11 Let $r = (1, 1, 3, 4)$, and let L denote the (nilpotent) Lie algebra of analytic vector fields on $\mathbb{R}^4$, which may be written as sums of δ^r−homogeneous vector fields of negative degree. Pick

$$f := \left(4x_1^2 + 4x_1 x_2 + x_2^2 - 2x_1 - x_2\right)\frac{\partial}{\partial x_3} + \left(x_1^3 - 3x_1^2 x_2 - 2x_3\right)\frac{\partial}{\partial x_4} \in L.$$

According to Proposition 5.9, there exists a (non-trivial) $s \in L$ such that $[s, f] = 0$. The set of such symmetries is easily found to be

$$\mathrm{Span}\{f, \frac{\partial}{\partial x_1} - 2\frac{\partial}{\partial x_2} + (\frac{9}{2}x_1^2 - 3x_1 x_2)\frac{\partial}{\partial x_3}\} \oplus \mathbb{R}_3[x_1, x_2]\frac{\partial}{\partial x_4}.$$

5.4.3 Symmetric Liapunov functions

We first show that Theorem 5.8 may be extended to any infinitesimal symmetry.

Theorem 5.13 *[117] Let f and s be two vector fields of class C^1 on $\mathbb{R}^n$. Assume that*
(i) s is a symmetry for f (i.e., $[s, f] = kf$ for some number k), the origin is globally AS for $\dot{x} = -s(x)$, and all solutions of this system are defined on $\mathbb{R}$;
(ii) The origin is globally AS for $\dot{x} = f(x)$.
Pick any $m > 0$. Then there exists a strict Liapunov function $V \in C^0(\mathbb{R}^n) \cap C^1(\mathbb{R}^n \setminus \{0\})$ which is s−homogeneous of degree m.

Proof. Let $\varphi(t, x)$ denote the flow of s. Pick any nondecreasing function $a : \mathbb{R}^+ \to [0, 1]$ of class C^∞, such that $a(t) = 0$ for $t \leq 1$, $a(t) = 1$ for $t \geq 2$

and $a' > 0$ on $(1, 2)$. According to Theorem 2.4, there exists a strict Liapunov function W in the large of class C^∞ for (5.1). We set

$$V(x) := \int_{-\infty}^{+\infty} e^{-mt}(a \circ W)(\varphi(t, x))\, dt, \quad \forall x \in \mathbb{R}^n.$$

Let a compact set $K \subset \mathbb{R}^n \setminus \{0\}$ be given. Clearly, there exist times $t_1 < t_2$ such that, for each $x \in K$, $W(\varphi(t, x)) \leq 1$ for all $t \leq t_1$ and $W(\varphi(t, x)) \geq 2$ for all $t \geq t_2$. Therefore

$$V(x) := \int_{t_1}^{t_2} e^{-mt}(a \circ W)(\varphi(t, x))\, dt + \frac{e^{-mt_2}}{m}, \quad \forall x \in K.$$

It follows that V is a (well-defined) function of class C^1 on $\mathbb{R}^n \setminus \{0\}$. V is clearly positive definite. On the other hand, for each pair $(T, x) \in \mathbb{R} \times \mathbb{R}^n$,

$$V(\varphi(T, x)) = \int_{-\infty}^{+\infty} e^{-mt}(a \circ W)(\varphi(t + T, x))\, dt = e^{mT} V(x),$$

thanks to an obvious change of integration variable. This shows that V is s-homogeneous of degree m, and continuous at the origin. It remains to prove that V is a strict Liapunov function for (5.1). For $x \neq 0$, we have

$$\langle \nabla V(x), f(x) \rangle = \int_{-\infty}^{+\infty} e^{-mt} a'\bigl(W(\varphi(t, x))\bigr) \langle \nabla W(\varphi(t, x)), \frac{\partial \varphi}{\partial x}(t, x) f(x) \rangle\, dt. \tag{5.58}$$

We claim that

$$f(\varphi(t, x)) = e^{kt} \frac{\partial \varphi}{\partial x}(t, x) f(x), \quad \forall (t, x) \in \mathbb{R} \times \mathbb{R}^n. \tag{5.59}$$

Indeed, if ψ is the flow of f, then (by Proposition 5.8) for small enough $\tilde{t}$

$$\psi(\tilde{t}, \varphi(t, x)) = \varphi(t, \psi(e^{kt}\tilde{t}, x)). \tag{5.60}$$

Differentiating (5.60) with respect to $\tilde{t}$ at $\tilde{t} = 0$ gives (5.59). Incorporating (5.59) into (5.58), we obtain

$$\langle \nabla V(x), f(x) \rangle = \int_{-\infty}^{+\infty} e^{-(m+k)t} a'\bigl(W(\varphi(t, x))\bigr) \langle \nabla W, f \rangle(\varphi(t, x))\, dt < 0.$$

∎

Remark 5.4 The above proof is suitable for Theorem 5.8, even if f is merely continuous.

Remark 5.5 Applying Theorem 5.13 with $s = -f$, we obtain a strict Liapunov function V (of class C^1 away from 0), such that

$$\langle \nabla V(x), f(x) \rangle = -mV(x).$$

It means that the inequality (4.8) in Theorem 4.5 may be changed into an equality.

Next result extends Hermes' theorem.

Corollary 5.6 *[117] Let f and s be as in Theorem 5.13. Let g be another vector field of class C^1 on $\mathbb{R}^n$, fulfilling*

$$\left(\frac{\partial \varphi}{\partial x}\right)^{-1} (\ln \varepsilon, x) g(\varphi(\ln \varepsilon, x)) = o(\varepsilon^k) \quad \text{as } \varepsilon \to 0, \tag{5.61}$$

uniformly with respect to x on S^{n-1}. Then the origin is still AS for (5.22).

Remark 5.6 When $s = \sum_i r_i x_i \frac{\partial}{\partial x_i}$, $\varphi(\ln \varepsilon, x) = \delta_\varepsilon^r x$, so that (5.61) is nothing else than (5.41).

Proof. Let V as given by Theorem 5.13. Differentiating $V(\varphi(t,x)) = e^{mt}V(x)$ with respect to x, we get

$$\frac{\partial V}{\partial x}(\varphi(t,x))\frac{\partial \varphi}{\partial x}(t,x) = e^{mt}\frac{\partial V}{\partial x}(x) \quad \forall t, \forall x \neq 0.$$

Hence, using (5.59),

$$\langle \nabla V(\varphi(\ln \varepsilon, x)), (f+g)(\varphi(\ln \varepsilon, x)) \rangle$$
$$= \varepsilon^m \langle \nabla V(x), \left(\frac{\partial \varphi}{\partial x}\right)^{-1}(\ln \varepsilon, x)\left(\varepsilon^k \frac{\partial \varphi}{\partial x}(\ln \varepsilon, x) f(x) + g(\varphi(\ln \varepsilon, x))\right)\rangle$$
$$= \varepsilon^{m+k}\left(\langle \nabla V(x), f(x)\rangle + \varepsilon^{-k}\langle \nabla V(x), \left(\frac{\partial \varphi}{\partial x}\right)^{-1}(\ln \varepsilon, x) g(\varphi(\ln \varepsilon, x))\rangle\right)$$
$$< 0,$$

for $x \in S^{n-1}$ and $\varepsilon > 0$ small enough (say, $0 < \varepsilon < \varepsilon_0$), thanks to (5.61). Therefore, V is also a strict Liapunov function for (5.22). ∎

Example 5.12 Let $f = -x_1^5 \frac{\partial}{\partial x_1} - x_2^3 \frac{\partial}{\partial x_2} - x_3 \frac{\partial}{\partial x_3}$ in $\mathbb{R}^3$, and $s = -f$. The above corollary cannot be applied, since s is not complete. Nevertheless, since we only are concerned with the behavior of trajectories around the origin, we may modify f and s (away from 0) in such a way that s becomes complete and the relation $[s, f] = 0$ still holds true. Let φ denote the flow of s. Then straightforward computations yield, for all x near the origin and all $t \leq 0$

$$\varphi(t, x) = \left((1 - 4tx_1^4)^{-\frac{1}{4}} x_1, (1 - 2tx_2^2)^{-\frac{1}{2}} x_2, e^t x_3\right),$$
$$\left(\frac{\partial \varphi}{\partial x}\right)^{-1}(t, x) = \mathrm{diag}\left((1 - 4tx_1^4)^{\frac{5}{4}}, (1 - 2tx_2^2)^{\frac{3}{2}}, e^{-t}\right).$$

The easy proof of next lemma is left to the reader.

Lemma 5.2 Let p, q, r, s be positive integers such that $\frac{p+1}{p} < \frac{s}{p} + \frac{r}{q}$. Then

$$(1 - pt|x|^p)^{\frac{p+1-s}{p}} |x|^s \left|(1 - qt|y|^q)^{-\frac{1}{q}} y\right|^r \to 0 \quad \text{as } t \to -\infty,$$

uniformly with respect to (x, y) on $\{(x, y) \in \mathbb{R}^2 : \|(x, y)\|_\infty = 1\}$.

With this lemma at hand, we readily infer from Corollary 5.6 that the origin is still AS for (5.22), provided that the components g_i of g fulfill

$$\begin{aligned}
g_1(x) &= O(x_1^6) + O(x_1^4 x_2) + O(x_1^2 x_2^2) + O(x_2^3) + O(x_3), \\
g_2(x) &= O(x_1^7) + O(x_1^5 x_2) + O(x_1^3 x_2^2) + O(x_1 x_2^3) + O(x_2^4) + O(x_3), \\
g_3(x) &= O(x_3^2) + O(x_1 x_3) + O(x_2 x_3).
\end{aligned}$$

For instance, 0 is AS for

$$\begin{cases}
\dot{x}_1 &= -x_1^5 + x_2^3, \\
\dot{x}_2 &= -x_2^3 + x_1^5 x_2 + x_1 x_3, \\
\dot{x}_3 &= -x_3 + x_2 x_3.
\end{cases}$$

■

Example 5.13 Let us consider the system

$$\begin{cases}
\dot{x}_1 &= -x_1^3 + x_2^2, \\
\dot{x}_2 &= -x_2.
\end{cases}$$

The origin is AS for this system, according to Corollary 5.6 and Lemma 5.2. (Take $f := -x_1^3 \frac{\partial}{\partial x_1} - x_2 \frac{\partial}{\partial x_2} =: -s$.) However, the same reasoning as in Example 5.9 shows that whatever the family of dilations and the new coordinates, the origin is no longer AS for the first homogeneous approximating system.

■

Example 5.14 Let us consider the single input control system

$$\begin{cases}
\dot{x}_1 &= -x_1 + x_1(x_2 - x_3), \\
\dot{x}_2 &= -x_2^3, \\
\dot{x}_3 &= u.
\end{cases}$$

Assume that we are interested in designing a feedback law u which yields the "best" rate of convergence for *only* the first coordinate x_1. The (natural) choice $u = -k\, x_3$ ($k > 0$ some constant) results in

$$x_1(t) = x_1(0) \exp[-t + \frac{2t x_2(0)}{(1 + 2t x_2(0)^2)^{\frac{1}{2}} + 1} - \frac{x_3(0)}{k}(1 - e^{-kt})],$$

hence $x_1(t)$ decreases like $e^{-t+\sqrt{2t}}$. On the other hand, the choice $u = -x_2^3 + x_2 - x_3$ (for which x_3 decreases like x_2), leads to

$$x_1(t) = x_1(0) \exp[-t + (x_2(0) - x_3(0))(1 - e^{-t})],$$

hence $x_1(t)$ decreases like e^{-t}. (See [108] for another application of symmetries to the stabilization problem.)

Next corollary extends a result in [51].

Corollary 5.7 *[117] Let f and s be as in Theorem 5.13. Let $\tilde{f}$ be a vector field on $\mathbb{R}^p$ such that $0 \in \mathbb{R}^p$ is AS for $\dot{y} = \tilde{f}(y)$. Let $h : \mathbb{R}_x^n \times \mathbb{R}_y^p \to \mathbb{R}^n$ be a continuous map such that*
(i) $0 \in \mathbb{R}^n$ *is globally AS for the system* $\dot{x} = f(x) + h(x,0)$;
(ii)
$$\left(\frac{\partial \varphi}{\partial x}\right)^{-1}(\ln \varepsilon, x)\, h(\varphi(\ln \varepsilon, x), y) = o(\varepsilon^k) \quad \text{as } \varepsilon \to +\infty,$$

uniformly with respect to (x,y) on each compact set in $\mathbb{R}^{n+p}$. Then $(0,0) \in \mathbb{R}^{n+p}$ is AS for the system

$$\begin{cases} \dot{x} = f(x) + h(x,y), \\ \dot{y} = \tilde{f}(y). \end{cases}$$

Proof. We proceed as in [51]. According to a result in [131], it is sufficient to prove that the system

$$\dot{x} = f(x) + h(x,u)$$

is BIBS-stable. Let V be as given in Theorem 5.13. We aim to prove that for each $S > 0$, there exists $R > 0$ such that

$$\langle \nabla V(x), f(x) + h(x,u) \rangle < 0 \quad \text{if } \|x\| > R,\ \|u\| < S.$$

The same computation as in the proof of Corollary 5.6 yields

$$\langle \nabla V(\varphi(\ln \varepsilon, x)), f(\varphi(\ln \varepsilon, x)) + h(\varphi(\ln \varepsilon, x), u) \rangle$$
$$= \varepsilon^{m+k}\{\langle \nabla V(x), f(x) \rangle + \varepsilon^{-k}\langle \nabla V(x), \left(\tfrac{\partial \varphi}{\partial x}\right)^{-1}(\ln \varepsilon, x) h(\varphi(\ln \varepsilon, x), u) \rangle\}$$
$$< 0,$$

if $\|x\| = 1$, $\|u\| \le S$ and ε is large enough (say, $\varepsilon > \varepsilon_0$). To complete the proof it is sufficient to observe that the set $\{\varphi(\ln \varepsilon, x);\ \varepsilon > \varepsilon_0,\ \|x\| = 1\}$ contains the complement of the closed ball B_R, provided that R is large enough. ■

Example 5.15 The system

$$\begin{cases} \dot{x}_1 &= -x_1^5 + |x_1 x_2|, \\ \dot{x}_2 &= -x_2 \end{cases}$$

is globally AS. (Apply Corollary 5.7 with $n = p = 1$ and $f = -x_1^5 \frac{\partial}{\partial x_1}$.)

Example 5.16 The system

$$\begin{cases} \dot{x}_1 &= x_2 + x_3^2 x_1^4, \\ \dot{x}_2 &= -x_1^9 - x_1^4 x_2 + x_3(x_2 x_1^3 + x_1^8), \\ \dot{x}_3 &= -x_3^7 \end{cases}$$

is globally AS. (Indeed, $f = x_2 \frac{\partial}{\partial x_1} - (x_1^9 + x_1^4 x_2) \frac{\partial}{\partial x_2}$ is $\delta^{(1,5)}$−homogeneous of degree 4.)

Next result shows that it is sometimes possible to construct a strict Liapunov function inheriting several symmetries of the system (namely, two commuting symmetries which look like a "rotation" and a "dilation", respectively).

Corollary 5.8 *[117] Let f and s be as in Theorem 5.13, and let r be a vector field of class C^1 on $\mathbb{R}^n$ such that $r(0) = 0$, $[r, f] = k'f$ (k' is some number), $[r, s] = 0$, and the flow θ associated with r is T−periodic (i.e., $\theta(t+T, x) = \theta(t, x)$ $\forall t, x$). Pick any $m > 0$. Then there exists a strict Liapunov function $V \in C^0(\mathbb{R}^n) \cap C^1(\mathbb{R}^n \setminus \{0\})$, which is r−homogeneous of degree 0, and s−homogeneous of degree m.*

Proof. Pick any strict Liapunov function W in the large of class C^∞ for (5.1), as given in Theorem 2.4, and set

$$V_1(x) := \int_0^T W(\theta(\tau, x)) \, d\tau.$$

Clearly, V_1 is a positive definite function of class C^1 on $\mathbb{R}^n$, which fulfills $V_1(\theta(t, x)) = V_1(x)$ $\forall t, x$. V_1 is also a strict Liapunov function for (5.1), since for any $x \neq 0$

$$\begin{aligned} \langle \nabla V_1(x), f(x) \rangle &= \int_0^T \langle \nabla W(\theta(\tau, x)), \frac{\partial \theta}{\partial x}(\tau, x) f(x) \rangle \, dx \\ &= \int_0^T e^{-k'\tau} \langle \nabla W, f \rangle (\theta(\tau, x)) \, d\tau \quad \text{(by (5.59))} \\ &< 0. \end{aligned}$$

Then we set

$$V(x) := \int_{-\infty}^{+\infty} e^{-m\tau} (a \circ V_1)(\varphi(\tau, x)) \, d\tau,$$

where a and φ are as in the proof of Theorem 5.13. According to this proof, V is a strict Liapunov function for (5.1), which is continuous and C^1 away from 0, and s-homogeneous of degree m. It remains to prove that V is still r-homogeneous of degree 0, i.e. that

$$V(\theta(t,x)) = V(x) \quad \forall t, x. \tag{5.62}$$

Since $[r, s] = 0$, the flows θ and φ commute, hence

$$\begin{aligned} V_1(\varphi(\tau, \theta(t,x))) &= V_1(\theta(t, \varphi(\tau, x))) \\ &= V_1(\varphi(\tau, x)), \end{aligned}$$

(since V_1 is r-homogeneous of degree 0), and (5.62) follows. ∎

Example 5.17 Consider the system

$$\begin{cases} \dot{x}_1 = (2x_1^2 + x_2^2)^2(-x_1 + x_2), \\ \dot{x}_2 = (2x_1^2 + x_2^2)^2(-2x_1 - x_2), \\ \dot{x}_3 = -x_3^3. \end{cases}$$

It is globally AS, as may be seen with $V(x) := \|x\|^2$ as strict Liapunov function. Let

$$f = (2x_1^2 + x_2^2)^2 \left((x_2 - x_1)\frac{\partial}{\partial x_1} - (2x_1 + x_2)\frac{\partial}{\partial x_2} \right) - x_3^3 \frac{\partial}{\partial x_3}$$

be the corresponding vector field. Then f is $\delta^{(1,1,2)}$–homogeneous of degree 4; that is, $[s, f] = 2f$ for $s := x_1\frac{\partial}{\partial x_1} + x_2\frac{\partial}{\partial x_2} + 2x_3\frac{\partial}{\partial x_3}$. Let $r := x_2\frac{\partial}{\partial x_1} - 2x_1\frac{\partial}{\partial x_2}$. The orbits of r are the ellipses $\{x = (x_1, x_2, x_3) \in \mathbb{R}^3 : 2x_1^2 + x_2^2 = c_1, x_3 = c_2\}$ with $c_1 \geq 0$, $c_2 \in \mathbb{R}$ some constants. Then $[r, f] = 0 = [r, s]$. According to Corollary 5.8, a strict Liapunov function W which is r-homogeneous of degree 0 and s-homogeneous of degree 4 does exist. One easily sees that

$$W(x) := (2x_1^2 + x_2^2)^2 + x_3^2$$

is convenient. ∎

Chapter 6

Monotonicity and generalized derivatives

This chapter is a short survey about the following subject. Let $V(t,x)$ be a real function, defined for $(t,x) \in \mathbb{R}^+ \times \mathbb{R}^n$. Let $\mathcal{S}$ be the set of all solutions of a system of ordinary differential equations or, more generally, of a differential inclusion

$$\dot{x} \in F(t,x) \qquad (6.1)$$

given on $\mathbb{R}^+ \times \mathbb{R}^n$. As illustrated in the previous chapters, one crucial step in stability theory reduces to find conditions which enable us to recognize whether the composite function $g(t) = V(t, x(t))$ is non-increasing on its domain, for each $x(\cdot) \in \mathcal{S}$. When this happens, we shall also say that V *is non-increasing along F*. As every student knows, if the function at hand is differentiable, monotonicity is trivially related to the sign of the derivative. Similar relations exist also in a nonsmooth context, but they are nontrivial and require the use of more general concepts.

The results presented in Sections 6.4 and 6.5 are essentially based on [13]. They present the advantage of a simple formulation, but more general and improved versions can be found in the literature (see [102], [32], [63]).

6.1 Tools from nonsmooth analysis

For reader's convenience, we report in this section some notions and tools of nonsmooth analysis which are needed in our exposition. Our approach is essentially based on generalized directional derivatives, but we shall also make use of generalized differentials and gradients.

Let $N \geq 1$ be any integer number (in the sequel, we will focus in particular on the cases $N = 1 + n$ and $N = n$). Let moreover $f(x) : \mathbb{R}^N \to \mathbb{R}$ be defined on an open subset Q of $\mathbb{R}^N$. For any choice of $x \in Q, w \in \mathbb{R}^N$ and $h \in \mathbb{R}$ ($h \neq 0$), we are interested in the difference quotient

$$\mathcal{R}(h, x, w) = \frac{f(x + hw) - f(x)}{h}.$$

As well known, the usual directional derivative at a fixed point $\bar{x} \in Q$ with respect to a fixed direction $\bar{w} \in \mathbb{R}^N$ is given by

$$Df(\bar{x}, \bar{w}) = \lim_{h \to 0} \mathcal{R}(h, \bar{x}, \bar{w})$$

provided that the limit exists and it is finite. When the existence of the limit is not guaranteed, certain notions of generalized derivatives may represent useful substitutes. The most popular type of generalized derivatives are the so-called *Dini derivatives*. The basic idea is as follows. To f, $\bar{x}$ and $\bar{w}$ we associate four elements of the extended real line $\mathbb{R} \cup \{\pm\infty\}$, denoted respectively by $\overline{D^+}f(\bar{x}, \bar{w})$, $\underline{D^+}f(\bar{x}, \bar{w})$, $\overline{D^-}f(\bar{x}, \bar{w})$, $\underline{D^-}f(\bar{x}, \bar{w})$. The former is defined as

$$\limsup_{h \to 0^+} \mathcal{R}(h, \bar{x}, \bar{w})$$

and the others are defined in similar way, taking the infimum instead of the supremum and the left limit instead of the right one, according to the notation.

This idea can be exploited further on. The clue for the developments we are interested in, is the following trivial, but crucial remark: $\mathcal{R}(h, x, w)$ depends not only on h, but also on x and w.

The *upper right contingent derivative* $\overline{D_K^+}f(\bar{x}, \bar{w})$ is defined as

$$\limsup_{\substack{h \to 0^+ \\ w \to \bar{w}}} \mathcal{R}(h, \bar{x}, w).$$

Analogously, one can define $\underline{D_K^+}f(\bar{x}, \bar{w})$, $\overline{D_K^-}f(\bar{x}, \bar{w})$, $\underline{D_K^-}f(\bar{x}, \bar{w})$. Contingent derivatives are in some way related to the so-called contingent cone, introduced by Bouligand in 1930 ([7], p. 190). Contingent derivatives and Dini derivatives of the same type differ, in general. However, they coincide (and are finite) when f is locally Lipschitz continuous. They coincide also when $N = 1$ and $w \neq 0$.

The following relations hold:
$$D_K^+ f(\bar{x},\bar{w}) = \underline{D_K^-}(-f)(\bar{x},-\bar{w}) = -\overline{D_K^-}f(\bar{x},-\bar{w}) = -\overline{D_K^+}(-f)(\bar{x},\bar{w}).$$

More recently, the *upper Clarke directional derivative* $\overline{D_C}f(\bar{x},\bar{w})$ appears in the context of nonsmooth optimization theory ([36]). It is defined as

$$\limsup_{\substack{h\to 0 \\ x\to \bar{x}}} \mathcal{R}(h,x,\bar{w})$$

(in this case we do not distinguish between right and left limits, since they always coincide). Similarly, we can define $\underline{D_C}f(\bar{x},\bar{w})$. Note that $\underline{D_C}f(\bar{x},\bar{w}) = -\overline{D_C}f(\bar{x},-\bar{w})$.

It is not difficult to verify that the map

$$w \mapsto \overline{D^+}f(\bar{x},w)$$

from $\mathbb{R}^N$ to $\mathbb{R}\cup\{\pm\infty\}$ is positively homogeneous. The same is true for any other type of generalized (Dini, contingent or Clarke, upper or lower, left or right) derivative. In addition, $w \mapsto \overline{D_C}f(\bar{x},w)$ is subadditive (and hence a convex function).

The *Clarke generalized gradient* of f at x is given by

$$\partial_C f(x) = \{p \in \mathbb{R}^N : \forall w \in \mathbb{R}^N \text{ one has } \underline{D_C}f(x,w) \leq p\cdot w \leq \overline{D_C}f(x,w)\}.$$

The set $\partial_C f(x)$ is convex for each $x \in Q$. Moreover, if f is locally Lipschitz continuous, then $\overline{D_C}f(x,w)$ is finite for each (x,w) and hence $\partial_C f(x)$ turns out to be compact. The upper Clarke derivative can be recovered from Clarke gradient. Indeed,

$$\overline{D_C}f(x,w) = \sup_{p\in\partial_C f(x)} p\cdot w$$

(and, in a similar way, $\underline{D_C}f(x,w) = \inf_{p\in\partial_C f(x)} p\cdot w$).

If f is locally Lipschitz continuous, by Rademacher's Theorem its gradient $\nabla f(x)$ exists almost everywhere. Let S be the subset of $\mathbb{R}^N$ where the gradient does not exist. Then, it is possible to characterize Clarke generalized gradient as:

$$\partial_C f(x) = \operatorname{co}\left\{\lim_{i\to\infty} \nabla f(x_i),\ x_i \to x,\ x_i \notin S\cup S_1\right\}$$

where S_1 is any subset of $\mathbb{R}^N$, with $\mu(S_1) = 0$.

By analogy with Clarke's theory, we associate with the contingent derivatives the following two sets:

$$\underline{\partial} f(x) = \{p \in \mathbb{R}^n : \overline{D_K^-} f(x,w) \leq p \cdot w \leq \underline{D_K^+} f(x,w), \ \forall w \in \mathbb{R}^n\} \quad (6.2)$$

and

$$\overline{\partial} f(x) = \{p \in \mathbb{R}^n : \overline{D_K^+} f(x,w) \leq p \cdot w \leq \underline{D_K^-} f(x,w), \ \forall w \in \mathbb{R}^n\} \ .$$

These sets are both convex and closed and may be empty. In addition, they are bounded provided that the contingent derivatives take finite values for each direction. If one of them contains two distinct elements, the other is necessarily empty.

Note that since the contingent derivatives are not convex functions, it is not possible in general to recover their values for arbitrary directions from $\overline{\partial} f(x)$ and $\underline{\partial} f(x)$.

It turns out (see [63]) that $\overline{\partial} f(x)$ and $\underline{\partial} f(x)$ coincide respectively with the so-called *generalized super* and *sub-differentials*. They can be defined in an independent way, by means of a suitable extension of the classical definition of Fréchet differential. More precisely, one has

$$\overline{\partial} f(x) = \{p \in \mathbb{R}^n : \limsup_{h \to 0} \frac{f(x+h) - f(x) - p \cdot h}{|h|} \leq 0\}$$

and

$$\underline{\partial} f(x) = \{p \in \mathbb{R}^n : \liminf_{h \to 0} \frac{f(x+h) - f(x) - p \cdot h}{|h|} \geq 0\} \ .$$

Using this representation, it is not difficult to see that if $\underline{\partial} f(x)$ and $\overline{\partial} f(x)$ are both nonempty, then they coincide with the singleton $\{\nabla f(x)\}$ and f is differentiable at x in the classical sense.

Clarke gradient and generalized differentials are related by $\overline{\partial} f(x) \cup \underline{\partial} f(x) \subseteq \partial_C f(x)$.

Beside generalized super and sub-differentials introduced above, we shall need also the notion of proximal differential. In analytic terms, the *proximal subdifferential* of f at x is the set of all vectors p which enjoy the following property. There exist $\sigma > 0$ and $\delta > 0$ such that for each z with $|z - x| < \delta$,

$$f(z) - f(x) \geq p \cdot (z - x) - \sigma |z - x|^2 \ .$$

The proximal subdifferential is denoted by $\underline{\partial}_P f(x)$. The proximal superdifferential $\overline{\partial}_P f(x)$ is defined in analogous way. Of course, one has $\underline{\partial}_P f(x) \subseteq \underline{\partial} f(x)$ and $\overline{\partial}_P f(x) \subseteq \overline{\partial} f(x)$. The sets $\underline{\partial}_P f(x)$ and $\overline{\partial}_P f(x)$ are convex but not necessarily closed. They may be both empty even if f admits a differential in the classical sense at x.

Other relationship among these types of generalized derivatives, gradients and differentials, and comments on their possible geometric interpretation can be found in [39], [40], [7], [8], [114].

6.2 Functions of one variable

The composite map $g(t) = V(t, x(t))$ we are interested in, is a function of one real variable. Hence, it is convenient to start by recalling some results about monotonicity of such functions ([101] p. 207, [121] p. 347).

Lemma 6.1 *Let $g(t) : I \to \mathbb{R}$ be absolutely continuous on each compact subinterval of I. Then, $g(t)$ is non-increasing on I if and only if $g'(t) \leq 0$ for a.e. $t \in I$.*

Note in particular that since the solutions of (6.1) are absolutely continuous, $V(t, x(t))$ meets the absolute continuity assumption if V is locally Lipschitz continuous.

Lemma 6.2 *Let I be an open interval and let $g(t) : I \to \mathbb{R}$ be continuous on I. Then, the following statements are equivalent:*
  (i) $g(t)$ is non-increasing on I
  (ii) $\overline{D^+} g(t) \leq 0$ for all $t \in I$
  (iii) $\overline{D^-} g(t) \leq 0$ for all $t \in I$
  (iv) $\underline{D^+} g(t) \leq 0$ for all $t \in I$
  (v) $\underline{D^-} g(t) \leq 0$ for all $t \in I$.

Remark 6.1 We remark that in the previous lemma the inequalities are required to hold for all t, and that the conclusion becomes false if "for all" is replaced by "almost everywhere" (as an example, consider a double Cantor stair). ∎

The situation is more involved in the semi-continuous case. Indeed, the roles played by Dini derivatives become now distinct.

Lemma 6.3 *Let I be an interval of the form $I = [a, b)$. Let $g(t) : I \to \mathbb{R}$ be lower semi-continuous on I. Then, the following statements are equivalent:*

(i) $g(t)$ *is non-increasing on* I
(ii) $\overline{D^+}g(t) \leq 0$ *for all* $t \in I$
(iii) $\underline{D^+}g(t) \leq 0$ *for all* $t \in I$.

In a similar way, the statement "the lower semi-continuous function $g(t)$ is non-decreasing" can be equivalently characterized by the inequalities $\overline{D^-}g(t) \geq 0$ or $\underline{D^-}g(t) \geq 0$ for all $t \in I$.

When $g(t)$ is upper semi-continuous, the roles are exchanged: indeed, $g(t)$ turns out to be non-increasing if and only if $\overline{D^-}g(t) \leq 0$ (or $\underline{D^-}g(t) \leq 0$) for all $t \in I$. Analogously, it is non-decreasing if and only if $\overline{D^+}g(t) \geq 0$ (or $\underline{D^+}g(t) \geq 0$) for all $t \in I$.

For the sake of completeness, we report a simple and direct proof of Lemma 6.3.

Proof of Lemma 6.3. We shall prove that (i) $\iff$ (ii), since the proof that (i) $\iff$ (iii) is completely analogous. In fact, the "(i) $\implies$ (ii)" part is trivial, so that we can limit ourselves to the "(i) $\impliedby$ (ii)" part. Our reasoning closely follows with some modifications the argument of [121], p. 347.

First of all we prove the following claim, where the basic assumption is strengthened.

<u>Claim:</u> if $\overline{D^+}g(t) < 0$ for all $t \in I$, then g is non-increasing.

By contradiction, assume that there exist $\alpha, \beta \in I$ such that

$$\alpha < \beta \quad \text{but} \quad g(\alpha) < g(\beta) \ .$$

Since g is lower semi-continuous, it has an absolute minimum on $[\alpha, \beta]$. Let

$$t_0 = \sup\{t \in [\alpha, \beta] : g(t) = \text{the absolute minimum}\} \ .$$

Because of lower semi-continuity, it is clear that the sup in the previous formula is actually a maximum. Moreover, since the absolute minimum is less than or equal to $g(\alpha) < g(\beta)$, we see that $t_0 < \beta$.

At this point it is clear that for h sufficiently small and positive,

$$\frac{g(t_0 + h) - g(t_0)}{h} > 0 \ .$$

Hence, $\overline{D^+}g(t_0) \geq 0$, which is impossible in force of the assumption.

In order to complete the proof, we consider now the auxiliary function $h(t) = g(t) - \varepsilon t$, where ε is positive and arbitrary. Since $-\varepsilon t$ is differentiable in the ordinary sense, we have

$$\overline{D^+}h(t) = \overline{D^+}g(t) - \varepsilon .$$

The assumption $\overline{D^+}g(t) \leq 0$ implies therefore $\overline{D^+}h(t) < 0$. According to the claim, we see that $h(t)$ is non-increasing and the conclusion follows since the choice of ε is arbitrary. ∎

6.3 Ordinary differential equations

We are now ready to address the monotonicity problem stated at the beginning of the chapter. The simplest case occurs when S coincides with the set of solutions of an autonomous system of ordinary differential equations

$$\dot{x} = f(x) \tag{6.3}$$

f being defined and continuous[1] for each $x \in \mathbb{R}^n$, and V does not depend on time (namely, $V(t,x) = V(x)$). If in addition V is defined and everywhere differentiable for $x \in \mathbb{R}^n$, then clearly the composite function $g(t) = V(x(t))$ is non-increasing for each solution $x(\cdot)$ if and only if

$$\nabla V(x) \cdot f(x) \leq 0$$

for each $x \in \mathbb{R}^n$. If we keep the continuity assumption about f but we relax the differentiability assumption about V the situation becomes more involved.

The following result is classical (see [121]).

Theorem 6.1 *Let f be continuous and let V be locally Lipschitz continuous. The following statements are equivalent:*

(i) for each $x \in \mathbb{R}^n$, $\underline{D^+}V(x, f(x)) \leq 0$
(ii) V is non-increasing along the solutions of (6.3)

Proof. (i) $\Longrightarrow$ (ii). Since V is Lipschitz continuous and the solutions are C^1, the composite function $g(t) = V(x(t))$ is absolutely continuous, so that its derivative (in the usual sense) exists for all $t \in I \setminus N$, where N is a set of zero measure.

---
[1] Note that if f is continuous, all the solutions of (6.3) are *classical* i.e., everywhere differentiable and with a continuous derivative.

Let $\bar{t} \in I \setminus N$. We have
$$g'(\bar{t}) = \lim_{h \to 0} \frac{V(x(\bar{t}+h)) - V(x(\bar{t}))}{h}.$$
To compute the limit, we observe that
$$\frac{V(x(\bar{t}+h)) - V(x(\bar{t}))}{h} = \frac{V(\bar{x} + \dot{x}(\bar{t})h + o(h)) - V(\bar{x})}{h}$$
$$= \frac{V(\bar{x} + hv) - V(\bar{x})}{h}$$
$$+ \frac{V(\bar{x} + hv + o(h)) - V(\bar{x} + hv)}{h} \quad (6.4)$$
where we set $\bar{x} = x(\bar{t})$ and $v = \dot{x}(\bar{t}) = f(\bar{x})$. Using the Lipschitz continuity of V, for some constant L and sufficiently small h we have
$$|V(\bar{x} + hv + o(h)) - V(\bar{x} + hv)| \le Lo(h).$$
This yields
$$\lim_{h \to 0^+} \frac{V(\bar{x} + hv + o(h)) - V(\bar{x} + hv)}{h} = 0.$$
We emphasize that the previous limit exists in the usual sense. Coming back to (6.4), and taking the lower right limit to both sides, we get
$$\underline{D^+} g(\bar{t}) = \underline{D^+} V(\bar{x}, v). \quad (6.5)$$
In addition, since $\bar{t} \notin N$, we have that $g'(\bar{t}) = \underline{D^+} g(\bar{t})$. Because of the assumption (i), we have so proved that
$$g'(t) \le 0$$
a.e. for $t \in I$. The conclusion follows from Lemma 6.1.

(i) $\Longleftarrow$ (ii). Let us fix $\bar{x} \in \mathbb{R}^n$; since f is continuous, there exists a solution $x(t)$ such that $x(0) = \bar{x}$. From the monotonicity condition, it follows that $\underline{D^+} g(0) \le 0$. By repeating the same computations as above, from (6.5) we conclude that
$$\underline{D^+} V(\bar{x}, f(\bar{x})) \le 0.$$
Since the choice of $\bar{x}$ is arbitrary, the proof is complete. ∎

We want now to generalize this result to the case where the function V is only semi-continuous.

Proposition 6.1 *Let V be lower semi-continuous and let f be continuous. Assume that*
$$\overline{D_K^+ V}(x, f(x)) \leq 0$$
for each $x \in \mathbb{R}^n$; then V is non-increasing along each solution of (6.3).

Proof. Let $g(t) = V(x(t))$, and recall that
$$\overline{D^+ g}(t) = \limsup_{h \to 0^+} \frac{V(x(t+h)) - V(x(t))}{h} .$$

Since $x(\cdot)$ is differentiable everywhere,
$$\overline{D^+ g}(t) = \limsup_{h \to 0^+} \frac{V(x(t) + h\dot{x}(t) + o(h)) - V(x(t))}{h} .$$

Set now $o(h) = \gamma(h)h$, where $\lim_{h \to 0} |\gamma(h)| = 0$, and $w(h) = \dot{x}(t) + \gamma(h)$. It is clear that $\lim_{h \to 0} w(h) = \dot{x}(t) = v = f(x(t))$. Hence

$$\begin{aligned}
\overline{D^+ g}(t) &= \limsup_{h \to 0^+} \frac{V(x(t) + hw(h)) - V(x(t))}{h} \\
&\leq \limsup_{\substack{h \to 0^+ \\ w \to v}} \frac{V(x(t) + hw) - V(x(t))}{h} \\
&= \overline{D_K^+ V}(x(t), v) .
\end{aligned}$$

By assumption, we see therefore that $\overline{D^+ g}(t) \leq 0$ for all $t \in I$ and the proof is completed by virtue of Lemma 6.3. ∎

Proposition 6.2 *Assume, as in Proposition 6.1, that f is continuous and that V is lower semi-continuous. Assume further that V is non-increasing along each solution of (6.3); then, $\underline{D_K^+ V}(x, f(x)) \leq 0$ for each $x \subset \mathbb{R}^n$.*

Proof. Let $x(t)$ be a solution such that $x(t) = x$, and let as before $\dot{x}(t) = v = f(x(t))$. Arguing as in the proof of Proposition 6.1, we see that

$$\begin{aligned}
\underline{D^+ g}(t) &= \liminf_{h \to 0^+} \frac{V(x(t+h)) - V(x(t))}{h} \\
&= \liminf_{h \to 0^+} \frac{V(x(t) + h\dot{x}(t) + o(h)) - V(x(t))}{h}
\end{aligned}$$

$$\geq \liminf_{\substack{h \to 0^+ \\ w \to v}} \frac{V(x(t)+hw) - V(x(t))}{h}$$

$$= D_K^+ V(x,v) .$$

Using again Lemma 6.3, since g is non-increasing, we have $D^+ g(t) \leq 0$. Hence, $\underline{D_K^+} V(x,v) \leq 0$. Since the choice of x is arbitrary, the proof is complete. ∎

Note that there is a gap between the statements of Propositions 6.1 and 6.2. In general, the situation cannot be improved.

Example 6.1 Inspired by a suggestion in [39], we consider a single valued (not Lipschitz continuous) equation

$$\dot{x} = \sqrt{|x|}$$

and a map

$$V(x) = \begin{cases} 0 & \text{if } x \leq 0 \\ 1 & \text{if } x > 0 \end{cases} .$$

Clearly V is increasing along the solution

$$x(t) = \begin{cases} -\frac{t^2}{4} & \text{if } t \leq 0 \\ \frac{t^2}{4} & \text{if } t \geq 0 \end{cases} .$$

Nevertheless, by a direct inspection we can see that $D_K^+ V(x, f(x)) \leq 0$ everywhere. This shows that the converse of Proposition 6.2 is false in general. To see that also the converse of Proposition 6.1 is false, we can use the same equation as before and the function

$$V(x) = \begin{cases} 1 & \text{if } x < 0 \\ 0 & \text{if } x \geq 0 \end{cases} .$$

The monotonicity property holds, but $\overline{D_K^+} V(0, f(0)) = \overline{D_K^+} V(0,0) = +\infty$. ∎

6.4 Differential inclusions

In this section we address the more general case where $\mathcal{S}$ coincides with the set of solutions of a differential inclusion of the form (6.1) that is, the set of all absolutely continuous functions $x(t) : I \to \mathbb{R}^n$ (where $I \subseteq \mathbb{R}^+$ is some interval) such that $\dot{x}(t) \in F(t, x(t))$ for a.e. $t \in I$. Our first goal is to generalize, as far as possible, Theorem 6.1. We use the shortened notation

$$\underline{D^+}V(t, x, v) \quad \text{instead of} \quad \underline{D^+}V((t, x), (1, v)) \,.$$

The following theorem is well known ([60], [54]). Its proof can be carried out as in the proof of Theorem 6.1, apart from some obvious modifications.

Theorem 6.2 *Let V be locally Lipschitz continuous on $\mathbb{R}^+ \times \mathbb{R}^n$, and let*

$$\sup_{v \in F(t,x)} \underline{D^+}V(t, x, v) \leq 0$$

for each $t \in \mathbb{R}^+$ and each $x \in \mathbb{R}^n$. Then, V is non-increasing along F.

Upper semi-continuity of set valued maps reduces to the usual notion of continuity in the single valued case. Nevertheless, Theorem 6.2 is not invertible not even if F fulfills assumptions ($\mathbf{H}_1$), ..., ($\mathbf{H}_4$) stated in Chapter 1. The reason is that in general F may contain "parasitic" directions: that is, for some $v_0 \in F(t_0, x_0)$ it may happen that there is no solution $x(t)$ such that $x(t_0) = x_0$ and $\dot{x}(t_0) = v_0$.

Example 6.2 We continue the study of system (4.86), already considered in Section 4.3 (Chapter 4). There, we noticed that the stability of the origin with respect to Filippov solutions can be checked by means of the Lipschitz continuous Liapunov function $V(x, y) = |x| + |y|$. This is clear from a geometrical point of view, but the following simple remark shows that Theorem 6.2 does not work in this case. Let $F(x, y)$ be the set valued map associated to (4.86) by Filippov's operator (1.3), and let $(x, y) = (1, 0)$. We have $v = (1, -3) \in F(1, 0)$, but $\underline{D^+}V((1, 0), (1, -3)) = 4$. For this example, the difficulty can be overcame by using, for instance, the method of [129] (see [12] for an improved version) based on Clarke gradient. ∎

A necessary and sufficient condition can be obtained under the additional assumption that F is continuous in the Hausdorff sense.

Theorem 6.3 *Assume that F is compact valued and continuous in the Hausdorff sense with respect to both t, x. A locally Lipschitz continuous function V is non-increasing along F if and only if*

$$\sup_{v \in F(t,x)} \underline{D^+V}(t,x,v) \leq 0$$

for each $t \in \mathbb{R}^+$ and each $x \in \mathbb{R}^n$.

Note that in the previous theorem, no convexity assumption is made about F. Preliminary to the proof of the theorem is the following proposition, which is a particular case of a result in [35] (see also [22]). A solution $x(t)$ of (6.1) is said to be *regular* if

- it is absolutely continuous
- its derivative is continuous, possibly except for a countable set of points, where only discontinuities of the first kind (jumps) may occur
- $\dot{x}(t) \in F(t, x(t))$ everywhere, possibly except for a countable set of points.

Proposition 6.3 *If the right hand side of (6.1) is compact valued and Hausdorff continuous with respect to both t, x, then for each t_0, x_0 and $v_0 \in F(t_0, x_0)$ there exists a regular solution such that $x(t_0) = x_0$ and $\dot{x}(t_0) = v_0$.*

The following proof of Theorem 6.3 is taken from [13].

Proof. The "if" part is nothing else than Theorem 6.2. As far as the "only if" part is concerned, since

$$\underline{D^+V}(t,x,v) \leq \overline{D_C V}(t,x,v) = \max\{p \cdot (1,v) : p \in \partial_C V(t,x)\}$$

for each t, x and $v \in F(t, x)$, it is sufficient to prove that

V non-increasing along $F \implies$
$p \cdot (1, v) \leq 0$ for each $p \in \partial_C V(t, x)$ and $v \in F(t, x)$.

Let us assume by contradiction that there exist $t \in \mathbb{R}^+$, $x \in \mathbb{R}^n$, $v \in F(t,x)$, $p \in \partial_C V(t, x)$ such that $p \cdot (1, v) > 0$. By the characterization of Clarke gradient, there exist $\lambda_i > 0$, $p_i \in \mathbb{R}^{n+1}$, $(i = 1, \ldots, m)$ such that $p = \sum_{i=1}^m \lambda_i p_i$, where:

a) $\sum_{i=1}^m \lambda_i = 1$,
b) there exist sequences $\{(t_k^i, x_k^i)\} \subseteq \mathbb{R}^+ \times \mathbb{R}^n$ such that:

- $\lim_{k \to +\infty} (t_k^i, x_k^i) = (t, x)$, $\forall i = 1, \ldots, m$
- V is differentiable at (t_k^i, x_k^i), and $\lim_{k \to +\infty} \nabla V(t_k^i, x_k^i) = p_i$, $\forall i = 1, \ldots, m$.

Since $p \cdot (1, v) = \sum_{i=1}^m \lambda_i (p_i \cdot (1, v)) > 0$, there exists at least one $j \in \{1, \ldots, m\}$ such that $p_j \cdot (1, v) > 0$.

Let us fix $\varepsilon < \min\left\{1, \frac{p_j \cdot (1,v)}{2(|p_j|+|v|+2)}\right\}$. Since $\lim_{k\to+\infty} \nabla V(t_k^j, x_k^j) = p_j$, there exists $\overline{k}$ such that

$$\forall k > \overline{k}, \exists w_k^j \in B_1 \subset \mathbb{R}^{n+1}, \quad \nabla V(t_k^j, x_k^j) = p_j + \varepsilon w_k^j.$$

By the continuity of F, there exists $\tilde{k}$ such that

$$\forall k > \tilde{k}, \exists v_k^j \in F(t_k^j, x_k^j), \exists z_k^j \in B_1 \subset \mathbb{R}^n, \quad v_k^j = v + \varepsilon z_k^j.$$

Then, $\forall k > \max\{\overline{k}, \tilde{k}\}$, we can write

$$\begin{aligned}
\nabla V(t_k^j, x_k^j) \cdot (1, v_k^j) &= (p_j + \varepsilon w_k^j) \cdot ((1,v) + \varepsilon(0, z_k^j)) \\
&\geq p_j \cdot (1,v) - \varepsilon |w_k^j \cdot (1,v) + (0, z_k^j) \cdot p_j + \varepsilon w_k^j \cdot (0, z_k^j)| \\
&\geq p_j \cdot (1,v) - \varepsilon(|v| + |p_j| + 2) > \frac{p_j \cdot (1,v)}{2} > 0.
\end{aligned}$$

Now we make use of Proposition 6.3. If we fix $K > \max\{\overline{k}, \tilde{k}\}$ and take $t_0 = t_K^j$, there exists a regular solution $x(t)$ such that $x(t_0) = x_K^j$ and $\dot{x}(t_0) = v_K^j$. Our choice of x_K^j allows us to conclude that $V(t, x(t))$ is differentiable at t_0, and:

$$\left.\frac{dV(t, x(t))}{dt}\right|_{t=t_0} = \nabla V(t_0, x(t_0)) \cdot (1, v_K^j) = \nabla V(t_K^j, x_K^j) \cdot (1, v_K^j) > 0,$$

a contradiction to the hypothesis of monotonicity of V along the solutions of (6.3). ∎

6.5 Monotonicity and the proximal gradient

If F is continuous in the Hausdorff sense but V is only semi-continuous, Theorem 6.3 is no more valid. Weakening the assumptions on V must be compensated by adding a new assumption on F. Note also that if V is no more locally Lipschitz continuous, $\underline{D^+}$ and D_K^+ differs: the right object to be considered in this setting is actually the contingent derivative.

Moreover, the investigation must be carried out by means of more powerful tools, such as proximal analysis.

For simplicity, in this section we limit ourselves to the time invariant case. First of all, we show that the condition of Theorem 6.3 can be reformulated in terms of proximal analysis.

Theorem 6.4 *Let F be compact valued and Hausdorff continuous, and let V be locally Lipschitz continuous. Then,*

$$\sup_{v \in F(x)} D^+V(x,v) \leq 0 \quad \forall x \in \mathbb{R}^n \tag{6.6}$$

if and only if

$$p \cdot v \leq 0 \quad \forall p \in \underline{\partial_P}V(x), \ \forall v \in F(x), \ \forall x \in \mathbb{R}^n . \tag{6.7}$$

Proof. The implication (6.6) $\longrightarrow$ (6.7) follows immediately from (6.2).

So, we prove that if (6.6) is false, then (6.7) is also false.

Let us therefore suppose that there exist $\bar{x} \in \mathbb{R}^n$, $\bar{y} \in F(\bar{x})$, $\eta > 0$ such that

$$D^+V(\bar{x}, \bar{y}) > \eta > 0.$$

Since V is Lipschitz continuous, $\underline{\partial_P}V(x)$ is locally bounded (see [40]), so that if we consider a compact neighborhood K of $\bar{x}$, there exists $M > 0$ such that

$$\forall x \in K, \ \forall p \in \underline{\partial_P}V(x): \quad |p| < M.$$

Since F is lower semicontinuous:

$$\exists \delta > 0 : |x - \bar{x}| < \delta \implies \exists y \in F(x), |y - \bar{y}| < \frac{\eta}{2M}. \tag{6.8}$$

We may take δ so that $\{x : |x - \bar{x}| < \delta\} \subseteq K$.

Since $D^+V(\bar{x}, \bar{y}) > \eta$, we may apply Subbotin's Theorem (see [40]). Therefore, there exist x, $|x - \bar{x}| < \delta$ and $q \in \underline{\partial_P}V(x)$ such that $q \cdot \bar{y} > \eta$. By (6.8), there exists $y \in F(x)$ such that $|y - \bar{y}| < \frac{\eta}{2M}$. Then:

$$|q \cdot (y - \bar{y})| \leq |q||y - \bar{y}| < M \cdot \frac{\eta}{2M} = \frac{\eta}{2}$$

and

$$q \cdot y = q \cdot \bar{y} + q \cdot (y - \bar{y}) > \eta - \frac{\eta}{2} = \frac{\eta}{2} > 0.$$

In such a way we have found that there exist x, $y \in F(x)$, $q \in \underline{\partial_P}V(x)$ such that $q \cdot y > 0$, thus contradicting (6.7). ∎

The following theorem, essentially well known, is a slight generalization of Theorem 7.2 in [39]. Its proof can be found in [13].

Theorem 6.5 *Let F be compact valued and locally Lipschitz continuous and let $V : \mathbb{R}^n \to \mathbb{R}$ be lower semicontinuous. Then, the following statements are equivalent:*

(i) V is non-increasing along all the solutions of (6.1).
(ii) $\sup \{p \cdot v \; : \; p \in \underline{\partial_P} V(x), v \in F(x)\} \leq 0, \qquad \forall x \in \mathbb{R}^n.$
(iii) $\sup_{v \in F(x)} \underline{D_K^+} V(x,v) \leq 0, \qquad \forall x \in \mathbb{R}^n.$

Bibliography

[1] Aeyels D., *Asymptotic Stability of Nonautonomous Systems by Liapunov's Direct Method*, Systems and Control Letters, **25** (1995), pp. 273-280

[2] Ancona F. and Bressan A., *Patchy Vector Fields and Asymptotic Stabilization*, ESAIM: Control, Optimisation and Calculus of Variations, **4** (1999), pp. 445-472

[3] Andriano V., Bacciotti A. and Beccari G., *Global Stability and External Stability of Dynamical Systems*, Nonlinear Analysis, Theory, Methods and Applications, **28** (1997), pp. 1167-1185

[4] Arnold V.I., *Algebraic Unsolvability of the Problem of Ljapunov Stability and the Problem of the Topological Classification of the Singular Points of an Analytic System of Differential Equations*, Functional Analalysis and its Applications, pp. 173-180 (translated from Funktsional'nyi Analiz i Ego Prilozheniya, **4** (1970), pp. 1-9)

[5] Artstein Z., *Stabilization with Relaxed Controls*, Nonlinear Analysis, Theory, Methods and Applications, **7** (1983), pp. 1163-1173

[6] Arzarello E. and Bacciotti A., *On Stability and Boundedness for Lipschitzian Differential Inclusions: the Converse of Liapunov's Theorems*, Set Valued Analysis, **5** (1998), pp. 377-390

[7] Aubin J.P. and Cellina A., *Differential Inclusions*, Springer Verlag, Berlin, 1984

[8] Aubin J.P. and Frankowska H., *Set Valued Analysis*, Birkhäuser, Boston, 1990

[9] Auslander J. and Seibert P., *Prolongations and Stability in Dynamical Systems*, Annales Institut Fourier, Grenoble, **14** (1964), pp. 237-268

[10] Bacciotti A., *Local Stabilizability of Nonlinear Control Systems*, World Scientific, Singapore, 1992

[11] Bacciotti A. and Beccari G., *External Stabilizability by Discontinuous Feedback*, Proceedings of the second Portuguese Conference on Automatic Control, 1996, pp. 495-498

[12] Bacciotti A. and Ceragioli F., *Stability and Stabilization of Discontinuous Systems and Nonsmooth Lyapunov Functions*, ESAIM: Control, Optimisation and Calculus of Variations, **4** (1999), pp. 361-376

[13] Bacciotti A., Ceragioli F., and Mazzi L., *Differential Inclusions and Monotonicity Conditions for Nonsmooth Liapunov Functions*, Set Valued Analysis, **8** (2000), pp. 299-309

[14] Bacciotti A. and Mazzi L., *Some Remarks on k-Asymptotic Stability*, Bollettino U.M.I. (7) 8-A (1994), pp. 353-363

[15] Bacciotti A. and Mazzi L., *A Necessary and Sufficient Condition for Bounded Input Bounded State Stability of Nonlinear Systems*, SIAM Journal on Control and Optimization, **39** (2000), pp. 478-491

[16] Bacciotti A. and Rosier L., *Liapunov and Lagrange Stability: Inverse Theorems for Discontinuous Systems*, Mathematics of Control, Signals and Systems, **11** (1998), pp. 101-128

[17] Bacciotti A. and Rosier L., *Regularity of Liapunov Functions for Stable Systems*, Systems and Control Letters, **41** (2000), pp. 265-270

[18] Bhat S.P. and Bernstein D.S., *Continuous Finite-Time Stabilization of the Translational and Rotational Double Integrators*, IEEE Transactions on Automatic Control, **43** (1998), pp. 678-682

[19] Bhat S.P. and Bernstein D.S., *Finite-Time Stability of Continuous Autonomous Systems*, SIAM Journal on Control and Optimization, **38** (2000), pp. 751-766.

[20] Bhatia N.P. and Szëgo G.P., *Stability Theory of Dynamical Systems*, Springer Verlag, Berlin, 1970

[21] Blagodatskikh V.I., *On the Differentiability of Solutions with respect to Initial Conditions*, Differential Equations, pp. 1640-1643 (translated from Differentsial'nye Uravneniya, **9** (1973), pp. 2136-2140)

[22] Blagodatskikh V.I. and Filippov A.F., *Differential Inclusions and Optimal Control*, In *Topology, Ordinary Differential Equations, Dynamical Systems*, Proceedings of Steklov Institute of Mathematics, 1986, pp. 199-259

[23] Bloch A. and Drakunov S., *Stabilization and Tracking in the Nonholonomic Integrator via Sliding Modes*, Systems and Control Letters, **29** (1996), pp. 91-99

[24] Bocharov A.V. et al., *Symmetries and Conservation Laws for Differential Equations of Mathematical Physics*, Translations of Mathematical Monographs **182**, American Mathematical Society, Providence, 1999

[25] Brezis H., *Analyse Fonctionnelle, Théorie et Applications*, Masson (1983)

[26] Brockett R., *Asymptotic Stability and Feedback Stabilization*, in *Differential Geometric Control Theory*, Ed.s Brockett R., Millman R., Sussmann H., Birkhäuser, Boston, 1983

[27] Byrnes C.I. and Isidori A., *New Results and Examples in Nonlinear feedback Stabilization*, Systems and Control Letters, **12** (1989), pp. 437-442

[28] Camilli F., Grüne L. and Wirth F., *A Generalization of Zubov's Method to Perturbed Systems*, SIAM Journal on Control and Optimization, **40** (2001), pp. 496-515

[29] Canudas de Witt C. and Sordalen O.J., *Examples of Piecewise Smooth Stabilization of Driftless NL Systems with less Input than States*, Proceedings of 2nd IFAC Symposium NOLCOS'92, Ed. Fliess M., Bordeaux, 1992, pp. 26-30

[30] Čelikovský S. and Nijmeijer H., *On the Relation Between Local Controllability and Stabilizability for a Class of Nonlinear Systems*, IEEE Transactions on Automatic Control, **42** (1997), pp. 90-94

[31] Ceragioli F., *Some Remarks on Stabilization by means of Discontinuous Feedback*, Systems and Control Letters, **45** (2002), pp. 271-281

[32] Cernea A. and Mirică Ş., *A Note on Nonsmooth Lyapunov Functions for State-Constrained Differential Inclusions*, Mathematical Reports (Romanian Academy), **5(55)** (2003), pp. 283-292

[33] Chiang H.D., Hirsch M.W. and Wu F.F., *Stability Regions of Nonlinear Autonomous Dynamical Systems*, IEEE Transactions on Automatic Control, **33** (1988), pp. 16-27

[34] Chiang H.D. and Thorp J.S., *Stability Regions of Nonlinear Dynamical Systems: a Constructive Methodology*, IEEE Transaction on Automatic Control, **34** (1989), pp. 1229-1241

[35] Chugunov P.I., *Regular Solution of Differential Inclusions*, Differential Equations, pp. 449-455, translated from Differentsial'nye Uravneniya, **17** (1981), pp. 660-668

[36] Clarke F.H., *Optimization and Nonsmooth Analysis*, Wiley and Sons, New York, 1983

[37] Clarke F.H., Ledyaev Yu.S., Sontag E.D. and Subbotin A.I., *Asymptotic Controllability Implies Feedback Stabilization*, IEEE Transactions on Automatic Control, **42** (1997), pp. 1394-1407

[38] Clarke F.H., Ledyaev Yu.S. and Stern R.J., *Asymptotic Stability and Smooth Lyapunov Functions*, Journal of Differential Equations, **149** (1998), pp. 69-114

[39] Clarke F.H., Ledyaev Yu.S., Stern R.J. and Wolenski P.R., *Qualitative Properties of Trajectories of Control Systems: a Survey*, Journal of Dynamical and Control Systems, **1** (1995), pp. 1-48

[40] Clarke F.H., Ledyaev Yu.S., Stern R.J. and Wolenski P.R., *Nonsmooth Analysis and Control Theory*, Springer Verlag, New York, 1998

[41] Coddington E.A. and Levinson N., *Theory of Ordinary Differential Equations*, McGraw-Hill, New York, 1955

[42] Conti R., *Linear Differential Equations and Control*, Academic Press, London, 1976

[43] Coron J.M., *Links between Local Controllability and Local Continuous Stabilization*, Proceedings of 2nd IFAC Symposium NOLCOS'92, Ed. Fliess M., Bordeaux, 1992, pp. 165-171

[44] Coron J.M., *Global Asymptotic Stabilization for Controllable Systems without Drift*, Mathematics of Control, Signals, and Systems, **5** (1992), pp. 295-312

[45] Coron J.M., *Stabilizing Time-varying Feedback*, Proceedings of 3rd IFAC Symposium NOLCOS'95, Ed.s A. Krener and D. Mayne, Tahoe, 1995

[46] Coron J.M., *Stabilization in Finite Time of Locally Controllable Systems by Means of Continuous Time-varying Feedback Law*, SIAM Journal on Control and Optimization, **33** (1995), pp. 804-833

[47] Coron J.M., *On the Stabilization of Some Nonlinear Control Systems: Results, Tools, and Applications* in Nonlinear Analysis, Differential Equations and Control, Ed.s Clarke F.H., Stern R.J., Kluwer, Dordrecht, 1999

[48] Coron J.M. and Praly L., *Adding an Integrator for the Stabilization Problem*, Systems and Control Letters, **17** (1991), pp. 89-104.

[49] Coron J.M. and Rosier L., *A Relation Between Continuous Time-Varying and Discontinuous Feedback Stabilization*, Journal of Mathematical Systems, Estimation and Control, **4** (1994), pp. 67-84

[50] Dayawansa W.P., *Recent Advances in the Stabilization Problem for Low-Dimensional Systems*, Proceedings of 2nd IFAC Symposium NOLCOS'92, Ed. Fliess M., Bordeaux, 1992, pp. 1-8

[51] Dayawansa W.P. and Martin C.F., *A Remark on a Theorem of Andreini, Bacciotti and Stefani*, Systems and Control Letters, **13** (1989), pp. 363-364

[52] Dayawansa W.P. and Martin C.F., *Asymptotic Stability of Nonlinear Systems with Holomorphic Structure*, Proceedings of 28th IEEE Conference on Decision and Control, Tampa, 1989

[53] Dayawansa W.P., Martin C.F. and Knowles G., *Asymptotic Stabilization of a Class of Smooth Two Dimensional Systems*, SIAM Journal on Control and Optimization, **28** (1990), pp. 1321-1349

[54] Deimling K., *Multivalued Differential Equations*, de Gruyter, Berlin, 1992

[55] Doob J.L., *Measure Theory*, Springer Verlag, New York, 1994

[56] Faubourg L. and Pomet J.B., *Control Lyapunov Functions for Homogeneous "Jurdjevic-Quinn" Systems*, ESAIM: Control, Optimisation and Calculus of Variations, **5** (2000), pp. 293-311

[57] Filippov A.F., *On Certain Questions in the Theory of Optimal Control*, SIAM Journal of Control, **1** (1962), pp. 76-84

[58] Filippov A.F., *Differential Equations with Discontinuous Right-hand Side*, Translations of American Mathematical Society, **42** (1964), pp. 199-231 (originally appeared on Matemat. Sbornik, **51** (1960), pp. 99-128)

[59] Filippov A.F., *Classical Solutions of Differential Equations with Multi-valued Right-Hand Side*, SIAM Journal of Control, **5** (1967), pp. 609-621

[60] Filippov A.F., *Differential Equations with Discontinuous Right-hand Side*, Kluwer Academic Publisher, Dordrecht, 1988

[61] Fradkov A.L., *Speed-Gradient Scheme and its Applications in Adaptive Control*, Automation and Remote Control, **40** (1979), pp. 1333-1342

[62] Frankowska H., *Hamilton-Jacobi Equations: Viscosity Solutions and Generalized Gradients*, Journal of Mathematical Analysis and Applications, **141** (1989), pp. 21-26

[63] Frankowska H., *Optimal Trajectories Associated with a Solution of the Contingent Hamilton-Jacobi Equation*, Applied Mathematics and Optimization, **19** (1989), pp. 291-311

[64] Galeotti M. and Gori F., *Bifurcations and Limit Cycles in a Family of Planar Polynomial Dynamical Systems*, Rendiconti del Seminario Matematico dell'Università e del Politecnico di Torino, **46** (1988), pp. 31-58

[65] Genesio R., Tartaglia M. and Vicino A., *On the Estimation of Asymptotic Stability Regions for Nonlinear Systems: State of the Art and New Proposals*, IEEE Transactions on Automatic Control, **30** (1985), pp. 747-755

[66] Hahn W., *Theory and Applications of Liapunov's Direct Method*, Prentice-Hall, Englewood Cliffs, 1963

[67] Hahn W., *Stability of Motions*, Springer Verlag, Berlin, 1967

[68] Haimo V.T., *Finite Time Controllers*, SIAM Journal on Control and Optimization, **24** (1986), pp. 760-770

[69] Hájek O., *Discontinuous Differential Equations, I*, Journal of Differential Equations, **32** (1979), pp. 149-170

[70] Hartman P., *Ordinary Differential Equations*, Birkhäuser, Boston, 1982

[71] Hautus M.L.J., *Stabilization, Controllability and Observability of Linear Autonomous Systems*, Indagationes Mathematicae, **32** (1970), pp. 448-455

[72] Hermes H., *The Generalized Differential Equation $\dot{x} \in R(t,x)$*, Advances in Mathematics, **4** (1970), pp. 149-169

[73] Hermes H., *Homogeneous Coordinates and Continuous Asymptotically Stabilizing Feedback Controls*, in: Differential Equations, Stability and Controls, Ed. S. Elaydi, Lecture Notes in Applied Math. **109**, Marcel Dekker, New York (1991), pp. 249-260

[74] Hermes H., *Nilpotent and High-Order Approximations of Vector Field Systems*, SIAM Review, **33** (1991), pp. 238-264

[75] Hong Y., Huang J. and Xu Y., *On an Output Feedback Finite-Time Stabilization Problem*, Proceedings of 38th IEEE Conference on Decision and Control, Phoenix, 1999, pp. 1302-1307

[76] Huang X. and Lin W., *Finite-Time Stabilization in the Large for Uncertain Nonlinear Systems*, preprint.

[77] Iggidr A., Kalitine B. and Outbib R., *Semi-Definite Lyapunov Functions: Stability and Stabilization*, Mathematics of Control, Signal, and Systems, **9** (1996), pp. 95-106

[78] Il'jašenko J.S., *Analytic Unsolvability of the Stability Problem and the Problem of Topological Classification of the Singular Points of Analytic Systems of Differential Equations*, Mathematics of the USSR-Sbornik, **28** (1976), pp. 140-152

[79] Isidori A., *Nonlinear Control Systems*, Springer Verlag, Berlin, 1989

[80] Jurdjevic V., *Geometric Control Theory*, Cambridge University Press, 1997

[81] Jurdjevic V. and Quinn J.P., *Controllability and Stability*, Journal of Differential Equations, **28** (1978), pp. 381-389

[82] Kalitine B., *Sur la stabilité des ensembles compact positivement invariants des systèmes dynamiques*, R.A.I.R.O. Automatique, **16** (1982), pp. 275-286

[83] Kawski M., *Nilpotent Lie Algebras of Vectorfields*, J. Reine Angew. Math., **388** (1988), pp. 1-17

[84] Kawski M., *Stabilization and Nilpotent Approximations*, Proceedings of 27th IEEE Conference on Decision and Control, 1988, pp. 1244-1248

[85] Kawski M., *Stabilization of Nonlinear Systems in the Plane*, Systems and Control Letters, **12** (1989), pp. 169-175

[86] Kawski M., *Homogeneous Stabilizing Feedback Laws*, Control Theory and Advanced Technology (Tokyo), **6** (1990), pp. 497-516

[87] Kawski M., *Geometric Homogeneity and Applications to Stabilization*, Proceedings of 3rd IFAC Symposium NOLCOS'95, Ed.s A. Krener and D. Mayne, Tahoe, 1995, pp. 147-152

[88] Krasowski N.N., *The Converse of the Theorem of K.P. Persidskij on Uniform Stability* (in russian), Prikladnaja Matematika I Mehanica, **19** (1955), pp. 273-278

[89] Krikorian R., *Necessary Conditions for a Holomorphic Dynamical System to Admit the Origin as a Local Attractor*, Systems and Control Letters, **20** (1993), pp. 315-318

[90] Kurzweil J., *On the Invertibility of the First Theorem of Lyapunov Concerning the Stability of Motion* (in russian with english summary), Czechoslovak Mathematical Journal, **80** (1955), pp. 382-398

[91] Kurzweil J., *On the Inversion of Liapunov's Second Theorem on Stability of Motion*, Translations of American Mathematical Society, **24** (1963), pp. 19-77 (originally appeared on Czechoslovak Mathematical Journal, **81** (1956), pp. 217-259)

[92] Kurzweil J. and Vrkoč I., *The Converse Theorems of Lyapunov and Persidskij Concerning the Stability of Motion* (in russian with english summary), Czechoslovak Mathematical Journal, **82** (1957), pp. 254-272

[93] La Salle J. and Lefschetz S., *Stability by Liapunov's Direct Method, with Applications*, Academic Press, New York, 1961

[94] Ledyaev Y.S. and Sontag E.D., *A Lyapunov Characterization of Robust Stabilization*, Nonlinear Analysis, Theory, Methods and Applications, **37** (1999), pp. 813-840

[95] Lin Y., Sontag E.D. and Wang Y., *A Smooth Converse Lyapunov Theorem for Robust Stability*, SIAM Journal on Control and Optimization, **34** (1996), pp. 124-160

[96] Malgrange B., *Ideals of Differentiable Functions*, Oxford University Press, Oxford, 1966

[97] Massera J.L., *On Lyapounoff's Conditions of Stability*, Annals of Mathematics, **50** (1949), pp. 705-721

[98] Massera J.L., *Contributions to Stability Theory*, Annals of Mathematics, **64** (1956), pp. 182-206

[99] M'Closkey R.T. and Murray R.M., *Non-holonomic Systems and Exponential Convergence: Some Analysis Tools*, Proceedings of 32nd IEEE Conference on Decision and Control, 1993, pp. 943-948

[100] M'Closkey R.T. and Murray R.M., *Exponential Stabilization of Driftless Nonlinear Control Systems Using Homogeneous Feedback*, IEEE Transactions on Automatic Control, **42** (1997), pp. 614-628

[101] McShane E.J., *Integration*, Princeton University Press, Princeton, 1947

[102] Mirică Ş., *Invariance and Monotonicity for Autonomous Differential Inclusions*, Studii si Cercetari Matematice, **47** (1995), pp. 179-204

[103] Morin P., Pomet J.B. and Samson C., *Design of Homogeneous Time-varying Stabilizing Control Laws for Driftless Controllable Systems via Oscillatory Approximation of Lie Brackets in Closed Loop*, SIAM Journal on Control and Optimization, **38** (1999), pp. 22-49

[104] Paden B.E. and Sastry S.S., *A Calculus for Computing Filippov's Differential Inclusions with Applications to the Variable Structure Control of Robot Manipulators*, IEEE Transactions on Circuits and Systems, **34** (1987), pp. 73-81

[105] A.Yu. Pogromsky, W.P.M.H. Heemels and H. Nijmeijer, *On Well Posedness of Relay Systems*, Proceedings of 5th IFAC Symposium NOLCOS'01, Elsevier Publ., 2002, pp. 1537-1542

[106] Pomet J.B., *Explicit Design of Time-varying Stabilizing Control Laws for a Class of Controllable Systems without Drift*, Systems and Control Letters, **18** (1992), pp. 147-158

[107] Pomet J.B. and Samson C., *Time-Varying Exponential Stabilization of Nonholonomic Systems in Power Form*, Technical Report 2126, INRIA, (1993)

[108] Praly L., *Generalized Weighted Homogeneity and State Dependent Time Scale for Linear Controllable Systems*, Proceedings of 36th IEEE Conference on Decision and Control, San Diego, 1997

[109] Prieur C., *A Robust Globally Asymptotically Stabilizing Feedback: the Example of the Artstein's Circles*, in "Nonlinear Control in the Year 2000" Ed.s Isidori A., Lamnabhi-Lagarrigue F. and Respondek W., Springer Verlag, 2000, pp. 279-300

[110] Qian C. and Lin W., *A Continuous Feedback Approach to Global Strong Stabilization of Nonlinear Systems*, IEEE Transactions on Automatic Control, **46** (2001), pp. 1061-1079

[111] Rifford L., *Stabilization des systèmes globalement asymptotiquement commandables*, Comptes Rendus de l'Académie des Sciences, Paris, Série I Mathématique, **330** (2000), pp. 211-216

[112] Rifford L., *Existence of Lipschitz and Semiconcave Control Lyapunov Functions*, SIAM Journal on Control and Optimization, **39** (2000), pp. 1043-1064

[113] Rifford L., *Nonsmooth Control-Lyapunov Functions; Application to the Integrator Problem*, preprint, Institut G. Desargues, Université Lyon I, France

[114] Rockafellar R.T. and Wets R.B., *Variational Analysis*, Springer Verlag, Berlin, 1998

[115] Rosier L., *Homogeneous Lyapunov Function for Homogeneous Continuous Vector Field*, Systems and Control Letters, **19** (1992), pp. 467-473

[116] Rosier L., *Inverse of Lyapunov's Second Theorem for Measurable Functions*, Proceedings of 2nd IFAC Symposium NOLCOS'92, Ed. Fliess M., Bordeaux, 1992, pp. 655-660

[117] Rosier L., *Etude de quelques Problèmes de Stabilisation*, Ph. D. Thesis, Ecole Normale Supérieure de Cachan (France), 1993.

[118] Rosier L., *Smooth Lyapunov Functions for Discontinuous Stable Systems*, Set-Valued Analysis, **7** (1999), pp. 375-405

[119] Rosier L. and Sontag E.D., *Remarks Regarding the Gap between Continuous, Lipschitz, and Differentiable Storage Functions for Dissipation Inequalities Appearing in H_∞ Control*, Systems and Control Letters, **41** (2000), pp. 237-249

[120] Rothschild L.P. and Stein E.M., *Hypoelliptic Differential Operators and Nilpotent Groups*, Acta Mathematica, **137** (1976), pp. 247-320

[121] Rouche N., Habets P. and Laloy M., *Stability Theory by Liapunov's Direct Method*, Springer Verlag, New York, 1977

[122] Rudin W., *Real and Complex Analysis*, McGraw Hill, London, 1970

[123] Rudin W., *Principles of Mathematical Analysis*, McGraw-Hill, New York, 1987.

[124] Ryan E.P., *On Brockett's Condition for Smooth Stabilizability and its Necessity in a Context of Nonsmooth Feedback*, SIAM Journal on Control and Optimization, **32** (1994), pp. 1597-1604

[125] Sansone C. and Conti R., *Nonlinear Differential Equations*, Pergamon, Oxford, 1964

[126] Sepulchre R. and Aeyels D., *Stabilizability does not Imply Homogeneous Stabilizability for Controllable Homogeneous Systems*, SIAM Journal on Control and Optimization, **34** (1996), pp. 1798-1813

[127] Sepulchre R. and Aeyels D., *Homogeneous Lyapunov Functions and Necessary Conditions for Stabilization*, Mathematics of Control, Signals, and Systems, **9** (1996), pp. 34-58

[128] Sepulchre R., Jankovic M. and Kokotovic P., *Constructive Nonlinear Control*, Springer Verlag, London, 1997

[129] Shevitz D. and Paden B., *Lyapunov Stability Theory of Nonsmooth Systems*, IEEE Transactions on Automatic Control, **39** (1994), pp. 1910-1914

[130] Sontag E.D., *A Lyapunov-like Characterization of Asymptotic Controllability*, SIAM Journal on Control and Optimization, **21** (1983), pp. 462-471

[131] Sontag E.D., *Smooth Stabilization Implies Coprime Factorization*, IEEE Transactions on Automatic Control, **34** (1989), pp. 435-443

[132] Sontag E.D., *A "Universal" Construction of Artstein's Theorem on Nonlinear Stabilization*, Systems and Control Letters, **13** (1989), pp. 117-123

[133] Sontag E.D., *Mathematical Control Theory*, Springer Verlag, New York, 1990

[134] Sontag E.D., *Feedback Stabilization of Nonlinear Systems*, in *Robust Control of Linear Systems and Nonlinear Control*, Ed.s Kaashoek M.A., van Schuppen J.H., Ran A.C.M., Birkhäuser 1990, pp. 61-81

[135] Sontag E.D., *On the Input-to-State Stability Property*, European Journal of Control, **1** (1995), pp. 24-36

[136] Sontag E.D., *Nonlinear Feedback Stabilization Revisited*, in *Dynamical Systems, Control, Coding, Computer Vision*, Ed.s Picci G., Gillian D.S., Birkhäuser, Basel, 1999, pp. 223-262

[137] Sontag E.D., *Stability and Stabilization: Discontinuities and the Effect of Disturbances*, in *Nonlinear Analysis, Differential Equations and Control*, Ed.s Clarke F.H., Stern R.J., Kluwer, Dordrecht, 1999

[138] Sontag E.D., *Clocks and Insensitivity to Small Measurement Errors*, ESAIM: Control, Optimisation and Calculus of Variations, **4** (1999), pp. 537-576

[139] Sontag E.D. and Sussmann H.J., *Remarks on Continuous Feedback*, Proceedings of 19th IEEE Conference on Decision and Control, Albuquerque, 1980, pp. 916-921

[140] Sontag E.D. and Sussmann H.J., *Further Comments on the Stabilizability of the Angular Velocity of a Rigid Body*, Systems and Control Letters, **12** (1989), pp. 213-217

[141] Sontag E.D. and Sussmann H.J., *Nonsmooth Control-Lyapunov Functions*, Proceedings of 34th IEEE Conference on Decision and Control, New Orleans, 1995, pp. 2799-2805

[142] Sontag E.D. and Wang Y., *On Characterizations of the Input-to-State Stability Property*, Systems and Control Letters, **24** (1995), pp. 351-359

[143] Sontag E.D. and Wang Y., *New Characterizations of Input-to-State Stability*, IEEE Transactions on Automatic Control, **41** (1996), pp. 1283-1294

[144] Sontag E.D. and Wang Y., *A Notion of Input to Output Stability*, Proceedings of European Control Conference, Bruxels 1997

[145] Sontag E.D. and Wang Y., *Notions of Input to Output Stability*, Systems and Control Letters, **38** (1999), pp. 235-248

[146] Sontag E.D. and Wang Y., *Lyapunov Characterizations of Input to Output Stability*, SIAM Journal on Control and Optimization, **39** (2000), pp. 226-249

[147] Sussmann H.J., *A General Theorem on Local Controllability*, SIAM Journal on Control and Optimization, **25** (1987), pp. 158-194

[148] Teel R.A. and Praly L., *A Smooth Lyapunov Function from a class.$\mathcal{KL}$ Estimate Involving Two Positive Semidefinite Functions*, ESAIM: Control, Optimisation and Calculus of Variations, **5** (2000), pp. 313-367

[149] Tsinias J., *Sufficient Lyapunov-Like Conditions for Stabilizability*, Mathematics of Control, Signals and Systems, **2** (1989), pp. 343-357

[150] Tsinias J., *A Local Stabilization Theorem for Interconnected Systems*, Systems and Control Letters, **18** (1992), pp. 429-434

[151] van der Schaft A., *L_2-gain and Passivity Techniques in Nonlinear Control*, Springer Verlag, London, 2000

[152] van der Schaft A., and Schumacher H., *An Introduction to Hybrid Dynamical Systems*, Lecture Notes in Control and Information Sciences 251, Springer Verlag, London 2000

[153] Vannelli A. and Vidyasagar M., *Maximal Lyapunov Functions and Domains of Attraction for Autonomous Nonlinear Systems*, Automatica, **21** (1985), pp. 69-80

[154] Varaiya P.P. and Liu R., *Bounded-input Bounded-output Stability of Nonlinear Time-varying Differential Systems*, SIAM Journal on Control, **4** (1966), pp. 698-704

[155] Vidyasagar M., *Nonlinear Systems Analysis*, Prentice Hall, Englewood Cliffs, 1993

[156] Wonham W.M., *Linear Multivariable Control: a Geometric Approach*, Springer Verlag, New York, 1979

[157] Yorke J.A., *Differential Inequalities and Non-Lipschitz Scalar Functions*, Mathematical Systems Theory, **4** (1970), pp. 140-153

Bibliography

[158] Yoshizawa T., *On the Stability of Solutions of a System of Differential Equations*, Memoirs of the College of Sciences, University of Kyoto, Ser. A, **29** (1955), pp. 27-33

[159] Yoshizawa T., *Liapunov's Functions and Boundedness of Solutions*, Funkcialaj Ekvacioj, **2** (1957), pp. 95-142

[160] Yoshizawa T., *Stability Theory by Liapunov's Second Method*, Publications of the Mathematical Society of Japan No. 9, 1966

[161] Zabczyk J., *Some Comment on Stabilizability*, Applied Mathematics and Optimization, **19** (1989), pp. 1-9

[162] Zabczyk J., *Mathematical Control Theory: an Introduction*, Birkhäuser, Boston, 1992

[163] Zubov V.I., *The Methods of Liapunov and their Applications*, Leningrad, 1957

Index

absolute stability, 39
affine system, 70
algebraic solvability, 167, 180
almost continuous stabilizer, 72
almost smooth stabilizer, 73
analytic Liapunov function, 169, 177
analytic solvability, 177
analytic vector field, 176
Arnold's Theorem, 176
Artstein's Example, 65
AS jet, 176
asymptotic controllability, 68
asymptotic stability, 28, 96
attraction, 95

BIBO-stability, 21
Borel measurability, 7

Carathéodory solution, 2
cascade system, 80
center-focus configuration, 32
Clarke derivative, 205
Clarke gradient, 205
Clarke, Ledyaev and Stern Theorem, 122
classical solution, 2
complete controllability, 21
contingent derivative, 204
continuous stabilizability, 62
control Liapunov function, 63

Coron's Theorem, 114, 115

damping control, 76
decay type, 169
derivative along an ODE, 29, 103
dilation, 181
Dini derivative, 204
discrete symmetry, 193
distribution, 80
domain of attraction, 49
double integrator, 190

equi-attraction, 96
equilibrium position, 95
essential continuity (ECT), 123
Euler vector field, 182
exponent, 170
exponential stability, 170

Filippov solution, 4, 5
finite gain property, 21
finite time stability, 175
finite time stabilization, 190
First Liapunov Theorem, 30, 104, 149

generalized differential, 206
generalized Liapunov function, 37

Hausdorff continuity, 8
Hermes' Theorem, 185, 198
holomorphic system, 180

homogeneous approximation, 168
homogeneous feedback, 187
homogeneous function, 181
homogeneous Liapunov function, 184
homogeneous norm, 183
homogeneous vector field, 181
Hurwitz matrix, 20, 170

Il'jašenko's Theorem, 177
impulse response, 21
infinitesimal symmetry, 193
infinitesimal upper bound, 103
input-to-output stability, 79
integrator, 81
Invariance Principle, 47, 93, 181, 191, 192
ISS-Liapunov function, 60
ISS-stability, 59
ISS-stabilizability, 62

jet, 176
Jurdjevic-Quinn method, 76

Kawski's Example, 63
Kawski's Theorem, 81
Krikorian's Theorem, 180
Kurzweil's Theorem, 31, 105

Lagrange stability, 27, 101
Lebesgue measurability, 7
Liapunov function, 29, 101
Liapunov stability, 27, 95
Lie bracket, 76
Lin, Sontag and Wang Theorem, 121
linear feedback, 22
local controllability (STLC), 68
Lojasewicz inequality, 178

maximal Liapunov function, 51
measurable set valued map, 6

neutral jet, 176
nonholonomic integrator, 65
nonholonomic system, 65

periodic feedback, 114
polynomial vector field, 168
positive definite function, 29, 103
positive limit set, 47
positive orbit, 40
prolongation, 40, 41

radially unbounded function, 29, 103
rational stability, 172
regular solution, 214
Riccati equation, 25
robust Lagrange stability, 112
robust stability, 108

Second Liapunov Theorem, 30, 104, 120
semianalytic set, 177
settling-time function, 175
simple function, 13
small control property, 72
stable matrix, 20
standard dilation, 182
strict Liapunov function, 29, 102
symmetric Liapunov function, 196
symmetry, 194

Taylor segment, 180
time varying feedback, 94, 113
total stability, 45
transitive map, 41
transitive prolongation, 42
triangular form, 81

Tsinias' Theorem, 81

UBIBS-Liapunov function, 59
UBIBS-stability, 58
UBIBS-stabilizability, 62
uniform global asymptotic stability
    (UGAS), 96, 120
uniform global attractivity, 96
uniform stability, 95
unstable jet, 176
upper semicontinuity, 6

weak Liapunov function, 29, 101
weight, 181
weighted homogeneity, 180

Yoshizawa's Theorem, 106

Zubov's equation, 50

List of abbreviations

AS	Asymptotic Stability
BIBO	Bounded-Input Bounded-Output
EA	Equi-Attraction
ECT	Essential Continuity with respect to Time
ISS	Input to State Stability
ODE	Ordinary Differential Equation
STLC	Small Time Local Controllability
UA	Uniform Attraction
UBIBS	Uniformly Bounded-Input Bounded-State
UGAS	Uniform Global Asymptotic Stability
US	Uniform Stability

Printing: Krips bv, Meppel
Binding: Litges & Dopf, Heppenheim